An Introduction to
the Theory of
Ordinary Differential Equations

An Introduction to the Theory of Ordinary Differential Equations

Walter Leighton

University of Missouri

Wadsworth Publishing Company, Inc., Belmont, California

Mathematics Editor: Don Dellen
Technical Illustrator: John Foster

ISBN 0–534–00432–6
L. C. Cat. Card No. 75–16957
Printed in the United States of America

1 2 3 4 5 6 7 8 9 10—
80 79 78 77 76

For Gin

Contents

Preface

Instead of preparing a fourth edition of the author's *Ordinary Differential Equations*, it appeared to be more appropriate to divide that book into two books—an elementary one and the present volume. This book is intended to be what its title suggests, an *introduction* to the theory of ordinary differential equations. It is designed for a one-semester course at the senior or first-year graduate level.

This is a textbook and not a treatise, which means, among other things, that the book is written primarily for the student. The material in it was selected either because of its central importance in the theory or because of its mathematical elegance—or ideally, both. The material was also selected to prepare the student to read more advanced books and papers in the field.

Experience at several universities indicates that it is likely that a number of students taking a first course in the theory of differential equations will have forgotten some elementary methods of solution. With this in mind, the book begins with what is believed to be a not-too-rapid review of formal methods (Chapters 1 and 2). A small amount of material from Chapter 2 is repeated in Chapter 4 for the convenience of students who start the course with Chapter 3.

Here and there in the book will be found a sprinkling of new material and some new methods. In this connection, special attention should be directed, perhaps, to Chapter 5 on oscillation, Chapter 8 on plane autonomous systems, Chapter 9 on Liapunov theory, and Chapter 10 on quadratic functionals.

A minimal amount of linear algebra is used in the book. Most students in a course in differential equations have some familiarity with at least the elements of linear algebra. My experience has been that the other students are not handicapped by taking on faith the small number of algebraic operations that I have introduced.

I have long been a believer in the principle that one learns mathematics by doing it, so a large number of exercises of varying degrees of difficulty have been provided. Most are designed to illuminate the theory.

My debt to others is very large. First of all, I should like to acknowledge my great debt to my teacher, Marston Morse. It was Professor Morse's brilliant and provocative lectures on differential equations that originally (and permanently) engaged my interest in the field. Next, I should point out that this and earlier editions of the book have had the benefit of searching comments by many colleagues. In particular, I am especially indebted to Courtney Coleman, Alfred Horn, Gerasimos E. Ladas, David A. Sanchez, and David V. V. Wend. While writing Chapter 9 on Liapunov theory I realized how much I owed not only to the books and papers of J. P. LaSalle, the late Solomon Lefschetz, and Jack K. Hale, but also to my many conversations with these men. Students and colleagues here at the University of Missouri have made many useful comments, and I wish to acknowledge my debt to them.

Finally, I should like to express the sincere hope that the teachers and students using this book will enjoy it. The area is an exciting one, and one's experience with it should not be somber.

Walter Leighton

An Introduction to
the Theory of
Ordinary Differential Equations

1

Review of elementary methods

1 Introduction

Differential equations are equations that involve derivatives. For example, the equations

$$y' = f(x),$$

$$y'' + y = 0,$$

(1.1) $$y'' = (1 + y'^2)^{1/2},$$

$$\frac{\partial^2 u}{\partial x^2} + \frac{\partial^2 u}{\partial y^2} = 0$$

are differential equations. The first three of these equations are called ordinary differential equations because they involve the ordinary derivatives of the unknown y. The last equation is an example of a partial differential equation, because it involves the partial derivatives of the unknown u. We shall be concerned with ordinary differential equations and their solutions.

To solve an algebraic equation, such as

(1.2) $$x^2 - 3x + 2 = 0,$$

we seek a number with the property that when the unknown x is replaced by this number the left-hand member of the equation reduces to zero. In equation (1.2) either the number 1 or the number 2 has this property. We say that this equation has the two solutions 1 and 2. To solve a differential equation we seek to determine not an unknown number but an unknown function. For example, in the equation

(1.3) $$y'' + y = 0,$$

y is regarded as the unknown. To find a solution we attempt to determine a function defined on an interval with the property that when y is replaced by this function, the equation reduces to an identity on this interval. It is clear that $\sin x$ is a solution of (1.3) for all values of x, for

$$(\sin x)'' + \sin x \equiv 0 \qquad (-\infty < x < \infty).$$

Similarly, it is easy to verify that $\cos x$ is also a solution of the differential equation (1.3).

Differential equations play a fundamental role in almost every branch of science and of engineering. They are of central importance in mathematical analysis. A differential equation describes the flow of current in a conductor; another describes the flow of heat in a slab. Other differential equations describe the motion of an intercontinental missile; still another describes the behavior of a chemical mixture. Sometimes it is important to find a particular solution of a given differential equation. Often we are more interested in the existence and behavior of solutions of a given differential equation than we are in finding its solutions.

In this chapter we shall begin our study by solving certain simple and important types of differential equations.

The *order* of a differential equation is the order of the highest derivative that appears in the equation. Accordingly, the first equation in (1.1) is of first order, and the next two equations are of second order. Similarly, the differential equation

$$y''' + y^4 = e^x$$

is of third order, and the equation

$$(y'''')^2 + yy' = 3$$

is of fourth order. The differential equation

(1.4) $$M(x, y) + N(x, y)y' = 0$$

is of first order. It is frequently useful to rewrite this equation in the form

(1.4)′ $$M(x, y)\, dx + N(x, y)\, dy = 0.$$

Thus,

$$(x^2 + y^2) + 2xy' = 0 \qquad \text{and} \qquad xe^y + (1 + y)y' = 0$$

are differential equations of first order which can be written, respectively, as

$$(x^2 + y^2)\, dx + 2x\, dy = 0 \quad \text{and} \quad xe^y\, dx + (1 + y)\, dy = 0.$$

Exercises

1. Find by inspection a solution of each of the following differential equations:
   (a) $y' - y = 0$;
   (b) $y' + 2y = 0$;
   (c) $y' = \sin x$.

2. Find by inspection a solution of each of the following differential equations:
   (a) $y'' - y = 0$;
   (b) $xy' - y = 0 \quad (x > 0)$;
   (c) $y'' = 0$.

2 Linear differential equations of first order

A linear differential equation of first order is an equation that can be put in the form

$$(2.1) \qquad\qquad k(x)y' + m(x)y = s(x).$$

On intervals on which $k(x) \neq 0$, both members of this equation may be divided by $k(x)$, and the resulting equation has the form

$$(2.2) \qquad\qquad y' + a(x)y = b(x).$$

We shall suppose that $a(x)$ and $b(x)$ are continuous on some interval I. The interval I may be any of the types $[a, b]$, (a, b), $[a, b)$, $(a, b]$, $(-\infty, b)$, $(-\infty, b]$, (a, ∞), $[a, \infty)$, or $(-\infty, \infty)$.

There are two commonly used elementary methods for solving an equation of the form (2.2).

Method 1. To solve equation (2.2), we may multiply both members of the equation by†

$$(2.3) \qquad\qquad e^{\int a(x)\, dx},$$

and we have‡

$$(2.4) \qquad\qquad [e^{\int a(x)\, dx}y]' = b(x)e^{\int a(x)\, dx}.$$

---

† By the symbol $\int a(x)\, dx$ is meant any function $A(x)$ such that $A'(x) = a(x)$ on I.
‡ Recall that $(e^{\int a(x)\, dx})' = a(x)e^{\int a(x)\, dx}$.

To solve (2.4) for y, we write

$$e^{\int a(x)\,dx} y = c + \int b(x) e^{\int a(x)\,dx}\,dx \qquad (c \text{ constant}),$$

and, finally,

(2.5) $$y = e^{-\int a(x)\,ax}\left[c + \int b(x) e^{\int a(x)\,dx}\,dx \right].$$

The student is advised to use the method described for solving an equation (2.2) rather than formula (2.5).

Example.   Solve the differential equation

(2.1)′ $$x^2 y' + xy = 2 + x^2 \qquad (x > 0).$$

We first put this equation in the form (2.2) by dividing it through by x^2:

(2.2)′ $$y' + \frac{1}{x} y = \frac{2}{x^2} + 1.$$

Here $a(x) = 1/x$ and $b(x) = 1 + 2/x^2$. We note that

(2.3)′ $$e^{\int a(x)\,dx} = e^{\int \frac{dx}{x}} = e^{\log x} = x.$$

If both members of (2.2)′ **are** multiplied by x, we have

(2.4)′ $$(xy)' = \frac{2}{x} + x.$$

From (2.4)′ we have successively

$$xy = 2\log x + \frac{x^2}{2} + c,$$

and

(2.5)′ $$y = \frac{2}{x}\log x + \frac{x}{2} + \frac{c}{x} \qquad (x > 0).$$

Method 2.   Consider the differential equation

(2.2) $$y' + a(x)y = b(x)$$

and the associated *homogeneous*† equation

---

† We shall follow the custom of *italicizing* words and phrases that are being defined either explicitly or implicitly in the text.

$$y' + a(x)y = 0.$$

It is easy to see by substitution that a solution of the latter equation is

$$e^{-\int a(x)\,dx}.$$

To complete the solution of (2.2) we introduce a new variable v in (2.2) by means of the substitution

(2.6)
$$y = e^{-\int a(x)\,dx}v,$$

$$y' = e^{-\int a(x)\,dx}[v' - a(x)v].$$

Equation (2.2) becomes

$$e^{-\int a(x)\,dx}v' = b(x),$$

and so

$$v' = b(x)e^{\int a(x)\,dx}.$$

The last equation yields

$$v = c + \int b(x)e^{\int a(x)\,dx}\,dx.$$

Using (2.6) we then have

$$y = e^{-\int a(x)\,dx}\left[c + \int b(x)e^{\int a(x)\,dx}\,dx\right],$$

which agrees with equation (2.5).

It will be instructive to apply the second method to the equation

(2.1)′
$$x^2y' + xy = 2 + x^2 \qquad (x > 0)$$

of the preceding example. The associated homogeneous equation

$$y' + \frac{1}{x}y = 0$$

has the solution

$$e^{-\int a(x)\,dx} = e^{-\int \frac{dx}{x}} = \frac{1}{x}.$$

We accordingly substitute

$$y = \frac{1}{x}v$$

in (2.1)′ obtaining

$$x^2\left(\frac{1}{x}v' - \frac{1}{x^2}v\right) + x\left(\frac{1}{x}v\right) = 2 + x^2 \qquad (x > 0),$$

or

$$v' = \frac{2}{x} + x.$$

Thus,

$$v = 2 \log x + \frac{x^2}{2} + c,$$

and

$$y = \frac{2}{x} \log x + \frac{x}{2} + \frac{c}{x} \qquad (x > 0),$$

which is (2.5)'.

The form of the solution of (2.2).   We have observed above that

$$e^{-\int a(x)\,dx}$$

is a solution of the homogeneous equation

(2.7) $y' + a(x)y = 0.$

Clearly, if c is any constant,

(2.8) $ce^{-\int a(x)\,dx}$

is also a solution. We shall learn later that (2.8) is the *general solution* of equation (2.7)—that is, every solution of (2.7) may be put in this form. Suppose now that $y_0(x)$ is some particular solution of the nonhomogeneous equation

(2.9) $y' + a(x)y = b(x);$

that is to say,

(2.10) $y_0'(x) + a(x)y_0(x) \equiv b(x).$

A substitution into (2.9) reveals that

(2.11) $y = y_0(x) + ce^{-\int a(x)\,dx}$

is then also a solution of (2.9) for each value of c. We shall see later that (2.11) provides the general solution of (2.9).

Example.   Consider the differential equation

(2.12) $y' + y = 3.$

The corresponding homogeneous equation is

$$y' + y = 0,$$

which has solutions

$$ce^{-x} \qquad (c \text{ constant}).$$

By inspection we note that $y = 3$ is a solution of (2.12). According to (2.11), the general solution of (2.12) is then given by

$$3 + ce^{-x}.$$

Exercises

Solve each of the following differential equations by two methods.

1. $y' + y = 3.$

2. $y' - 2y = x.$

3. $xy' + y = 2x + e^x.$

4. $x^2 y' - xy = x^3 + 4.$

5. $y' + ay = b \qquad (a, b \text{ constants}).$

6. $y' + \dfrac{1}{x} y = e^{x^2}.$

Solve the following differential equations.

7. $y'' + y' = 2.$ (*Hint.* Set $p = y'$.)

8. $xy'' + y' = x - 2.$

9. $y^2 \, dx + (y^2 x + 2xy - 1) \, dy = 0.$ (*Hint.* Reverse the roles of x and y.)

10. $y + (x - y^3 - 2)y' = 0.$

In the next four exercises find solutions $y(x)$ of the given differential equations which satisfy the given conditions.

11. $y' - y = 1, y(0) = 0.$

12. $xy' + y = 2x, y(1) = 1.$

13. $e^{-x}y' + 2e^x y = e^x, y(0) = \dfrac{1}{2} + \dfrac{1}{e}.$

14. $(\sin x)y' + (\cos x)y = \cos 2x, \; y\!\left(\dfrac{\pi}{2}\right) = \dfrac{1}{2}.$

15. Show that the substitution

$$u = y^{1-n}$$

reduces the so-called *Bernoulli* equation

$$y' + p(x)y = q(x)y^n \qquad (n \neq 1)$$

to the linear equation

$$\frac{1}{1 - n} u' + p(x)u = q(x).$$

16. Solve the Bernoulli equation
$$y' - y = xy^{1/2}.$$

17. Solve the Bernoulli equation
$$\frac{dx}{dy} + x = x^2.$$

18. Show that if $y_0(x)$ is a solution of the equation

$$y' + a(x)y = b(x),$$

then

$$y_0(x) + ce^{-\int a(x)\,dx} \qquad (c \text{ constant})$$

is also a solution of this equation for each value of the constant c.

Answers

1. $y = 3 + ce^{-x}.$

3. $y = x + \dfrac{1}{x} e^x + \dfrac{c}{x}.$

5. $y = \dfrac{b}{a} + ce^{-ax}.$

7. $y = 2x + c_1 e^{-x} + c_2.$

9. $x = y^{-2}[1 + ce^{-y}].$

11. $e^x - 1.$

13. $e^{-e^{2x}} + \frac{1}{2}.$

17. $x = (1 + ce^y)^{-1}.$

3 Exact differential equations of first order

A particularly important class of differential equations are the so-called *exact* differential equations. A differential equation

(3.1) $$M(x, y)\, dx + N(x, y)\, dy = 0$$

is said to be exact if there exists a function $g(x, y)$ such that

$$d[g(x, y)] = M(x, y)\, dx + N(x, y)\, dy;$$

that is to say, if there exists a function $g(x, y)$ such that

$$g_x(x, y) = M(x, y) \quad \text{and} \quad g_y(x, y) = N(x, y).$$

Thus, the equation

(3.2) $$(4x - y)\, dx + (2y - x)\, dy = 0$$

is exact, since its left-hand member is the differential of the function

$$g(x, y) = 2x^2 - xy + y^2,$$

for

$$d(2x^2 - xy + y^2) = (4x - y)\, dx + (2y - x)\, dy.$$

Clearly, we might equally well have chosen $g(x, y) = 2x^2 - xy + y^2 + 3$, or $g(x, y) = 2x^2 - xy + y^2 + c$, where c is any constant.

When $g(x, y)$ is a differentiable function such that

$$d[g(x, y)] = M(x, y)\, dx + N(x, y)\, dy,$$

any function $g(x, y) - c$, where c is a constant, is called an *integral* of the corresponding differential equation (3.1). Curves defined by the equations

$$g(x, y) = c \qquad (c \text{ constant})$$

are called *integral curves* of the differential equation. Accordingly, the function $2x^2 - xy + y^2$ is an integral of equation (3.2). Integral curves of equation (3.2) are given by the equation

(3.3) $$2x^2 - xy + y^2 = c.$$

When $c > 0$ the curves given by (3.3) are readily seen to be ellipses.

It is natural to inquire how we may identify those differential equations (3.1) that are exact, and how, when they are exact, corresponding integrals $g(x, y)$ may be determined. The following theorem is fundamental.

Theorem 3.1. *If the functions $M(x, y)$, $N(x, y)$ and the partial derivatives $M_y(x, y)$, $N_x(x, y)$ are continuous in a square region R, a necessary and sufficient condition that the differential equation*

$$M(x, y)\, dx + N(x, y)\, dy = 0$$

be exact is that

(3.4) $$\frac{\partial M}{\partial y} \equiv \frac{\partial N}{\partial x}.$$

The proof of the necessity of the condition is immediate. We suppose that the differential equation is exact; that is, there exists a function $g(x, y)$ with the property that

$$g_x(x, y) = M(x, y), \qquad g_y(x, y) = N(x, y).$$

Since $g_{xy} = g_{yx}$ it follows at once that

$$\frac{\partial M}{\partial y} = \frac{\partial N}{\partial x}.$$

The proof of the necessity is complete.

In proving the sufficiency we exhibit a function $g(x, y)$ whose partial derivatives satisfy the condition

$$g_x(x, y) = M(x, y) \qquad g_y(x, y) = N(x, y).$$

Such a function is†

$$(3.5) \qquad g(x, y) = \int_{x_0}^{x} M(x, y_0)\, dx + \int_{y_0}^{y} N(x, y)\, dy,$$

where (x_0, y_0) is a fixed point and (x, y) is an arbitrary point of the region R. For,

$$g_x = M(x, y_0) + \int_{y_0}^{y} N_x(x, y)\, dy$$

$$= M(x, y_0) + \int_{y_0}^{y} M_y(x, y)\, dy$$

$$= M(x, y_0) + [M(x, y) - M(x, y_0)]$$

$$= M(x, y),$$

while

$$g_y = N(x, y).$$

The proof of the theorem is complete.

Example.   In the differential equation

$$(3.6) \qquad (3x^2 + y^2)\, dx + 2xy\, dy = 0$$

we see that $M(x, y) = 3x^2 + y^2$, $N(x, y) = 2xy$, and $M_y = 2y$, $N_x = 2y$. Thus, the differential equation is exact. To find an integral $g(x, y)$ we choose the point (x_0, y_0) to be the origin, and we have

$$g(x, y) = \int_{0}^{x} (3x^2 + 0^2)\, dx + \int_{0}^{y} 2xy\, dy$$

$$(3.7) \qquad\qquad = x^3 + xy^2.$$

---

† The student who is familiar with line integrals will recognize the integral in (3.5) as the line integral $\int M\, dx + N\, dy$ taken over an "elbow path" from the point (x_0, y_0) to (x, y). Condition (3.4) will be seen to be precisely the condition that the line integral be independent of the path in R.

It is easily seen that the differential of (3.7) is given by the left-hand member of (3.6). Integral curves are given by the equation

(3.8) $x^3 + xy^2 = c,$

where c is a constant.

By finding (3.8) we have solved equation (3.6) in the sense that if we solve equation (3.8) for y and obtain a differentiable function of x, then that function is a solution of the differential equation. Specifically, we find that

$$y = \pm \sqrt{\frac{1}{x}(c - x^3)}.$$

It is easy to verify that both $\sqrt{\frac{1}{x}(c - x^3)}$ and $-\sqrt{\frac{1}{x}(c - x^3)}$ are indeed

solutions of (3.6) over a suitable interval of the x-axis.

This is the general situation, as will be seen from the following theorem.

Theorem 3.2. *If $g(x, y)$ is an integral of an exact differential equation $M(x, y)\,dx + N(x, y)\,dy = 0$, any differentiable solution $y(x)$ of the equation $g(x, y) = c$ is a solution of the differential equation.*

To prove the theorem note that because $g(x, y)$ is an integral of the differential equation it follows that

$$g_x(x, y) \equiv M(x, y), \qquad g_y(x, y) \equiv N(x, y).$$

Thus,

$$g_x[x, y(x)] \equiv M[x, y(x)], \qquad g_y[x, y(x)] \equiv N[x, y(x)].$$

Further, since $y(x)$ is a solution of the equation $g(x, y) = c$, we have

$$g[x, y(x)] \equiv c \qquad (c \text{ constant}).$$

It follows after differentiating this last identity that

$$g_x[x, y(x)] + g_y[x, y(x)]y'(x) \equiv 0,$$

and, hence,

$$M[x, y(x)] + N[x, y(x)]y'(x) \equiv 0;$$

that is to say, $y(x)$ is a solution of the given differential equation.

Alternate method. The line integral (3.5) provides a simple and direct method of solving an exact differential equation. An alternate method, the validity of which may be established by the preceding analysis, will be illustrated by an example.

We have observed that the differential equation

(3.9) $(3x^2 + y^2)\, dx + 2xy\, dy = 0$

is exact. To find an integral we first integrate the term $2xy\, dy$ formally with respect to y, obtaining

$$xy^2.$$

Next, we determine a function $f(x)$, of x alone, such that

$$d[xy^2 + f(x)]$$

is given by the left-hand member of (3.9). That is, we wish to find a function $f(x)$ such that

$$2xy\, dy + y^2\, dx + f'(x)\, dx = (3x^2 + y^2)\, dx + 2xy\, dy.$$

This is equivalent to the equation

$$f'(x) = 3x^2.$$

It follows that

$$f(x) = x^3 + c,$$

and that integrals of (3.9) are given by

$$xy^2 + x^3 + c.$$

We might equally well have commenced by integrating formally the term $(3x^2 + y^2)\, dx$, obtaining $x^3 + xy^2$. We would then seek to determine a function $g(y)$, of y alone, such that $d[x^3 + xy^2 + g(y)]$ is given by the left-hand member of (3.9).

An advantage of the alternate method may be observed in the first treatment of equation (3.9). Clearly, if we can determine the function $f(x)$, the equation is necessarily solved. It is desirable, however, to demonstrate that under the conditions of Theorem 3.1, such a function $f(x)$ can always be determined. This can be seen as follows. Consider the differential equation

$$M(x, y)\, dx + N(x, y)\, dy = 0,$$

and suppose that the conditions of Theorem 3.1 are satisfied. By "formal integration" of the term $N(x, y)\, dy$ is meant determining a function

$$H(x, y) = k + \int_{y_0}^{y} N(x, y)\, dy \qquad (k \text{ constant}),$$

where the points (x, y_0) and (x, y) lie in R. We note that

$$H_y(x, y) = N(x, y),$$

$$(3.10) \qquad H_x(x, y) = \int_{y_0}^{y} N_x(x, y)\, dy = \int_{y_0}^{y} M_y(x, y)\, dy$$

$$= M(x, y)\Big|_{y=y_0}^{y=y} = M(x, y) - M(x, y_0).$$

To complete the demonstration we show that there exists a function $f(x)$, of x alone, such that

$$(3.11) \qquad d[H(x, y) + f(x)] = M(x, y)\, dx + N(x, y)\, dy.$$

This is equivalent to demonstrating that there exists a function $f(x)$ such that

$$H_x(x, y)\, dx + H_y(x, y)\, dy + f'(x)\, dx = M(x, y)\, dx + N(x, y)\, dy,$$

or, by (3.10), such that

$$f'(x) = M(x, y_0).$$

It is clear that $f(x)$ may be taken as

$$f(x) = \int_{x_0}^{x} M(x, y_0)\, dx,$$

if (x_0, y_0) lies in R.

Remark. It is frequently desirable in differential equation theory to note a distinction between solving a differential equation and finding a solution of a differential equation. Recall that a *solution* is always a function of x, defined on an interval, which satisfies the differential equation. On the other hand, it is customary to regard a first-order differential equation as *solved* when one can write equations of its integral curves. Theorem 3.2, of course, justifies this seeming ambiguity.

Example. Find an integral curve of the differential equation

$$(3x^2 + y^2)\, dx + 2xy\, dy = 0$$

that passes through the point $(1, -2)$.

We saw earlier that integral curves of this differential equation are given by the equation

$$x^3 + xy^2 = c.$$

If we set $x = 1$, $y = -2$ in this equation, we find that $c = 5$; accordingly, the integral curve we seek is given by the equation

$$x^3 + xy^2 = 5.$$

Exercises

Show that the following differential equations are exact, and solve them. In the first five exercises sketch several integral curves.

1. $3x^2y \, dx + x^3 \, dy = 0$. Solve also as a linear differential equation.

2. $3(x - 1)^2 \, dx - 2y \, dy = 0$.

3. $(2x - y) \, dx + (2y - x) \, dy = 0$.

4. $(2x - y) \, dx - x \, dy = 0$. Solve also as a linear differential equation.

5. $(x - 2y) \, dx + (4y - 2x) \, dy = 0$.

6. $\cos x \sec y \, dx + \sin x \sin y \sec^2 y \, dy = 0$.

7. $\left(x + \dfrac{y}{x^2 + y^2}\right) dx + \left(y - \dfrac{x}{x^2 + y^2}\right) dy = 0$.

8. $(y^2 + 6x^2y) \, dx + (2xy + 2x^3) \, dy = 0$.

9. $(2xy + e^y) \, dx + (x^2 + xe^y) \, dy = 0$.

10. $(2x \cos y - e^x) \, dx - x^2 \sin y \, dy = 0$.

Find the solutions $y(x)$ of the following three differential equations that satisfy the given condition.

11. $(x^2 + y^2) \, dx + 2xy \, dy = 0$, $y(1) = 1$.

12. $\dfrac{y \, dx}{x^2 + y^2} - \dfrac{x \, dy}{x^2 + y^2} = 0$, $y(2) = 2$.

13. $(x - y) \, dx + (2y - x) \, dy = 0$, $y(0) = 1$.

14. Solve the differential equation (3.6) by evaluating the line integral $\int M \, dx + N \, dy$ along a straight line joining the points $(0, 0)$ and (x, y).

15. Do the same as Exercise 14 for the general differential equation of Theorem 3.1.

Answers

1. $x^3y = c$.

3. $x^2 - xy + y^2 = c^2$.

5. $x - 2y = c$.

9. $x^2y + xe^y = c$.

11. $[(4 - x^3)/3x]^{1/2}$.

13. $\frac{1}{2}(x + \sqrt{4 - x^2})$.

4 Variables separable

In the last section we learned how to treat an exact differential equation of the form

(4.1) $$M(x, y)\, dx + N(x, y)\, dy = 0.$$

When an equation of the form (4.1) is not exact we may try to find an *integrating factor*; that is, a function $I(x, y)$ with the property that the equation

(4.1)′ $$I(x, y)[M(x, y)\, dx + N(x, y)\, dy] = 0$$

is exact. In general, such a function exists but, except in certain special cases, it is likely to be difficult to determine. We limit ourselves to the case when the variables are separable.

When a differential equation has the form

(4.2) $$a(x)k(y)\, dx + b(x)h(y)\, dy = 0,$$

the variables are said to be *separable*. It is easy to see that

$$\frac{1}{b(x)k(y)}$$

is an integrating factor in regions of the xy-plane in which $b(x)k(y) \neq 0$. The differential equation (4.2) becomes

$$\frac{a(x)}{b(x)}\, dx + \frac{h(y)}{k(y)}\, dy = 0,$$

which is evidently exact. Integral curves are given by

$$\int_{x_0}^{x} \frac{a(x)}{b(x)}\, dx + \int_{y_0}^{y} \frac{h(y)}{k(y)}\, dy = C.$$

These integrals will exist if, for example, the integrands are continuous on the intervals of integration.

Example. Consider the differential equation

(4.3) $$x\, dy - y\, dx = 0.$$

It is easy to verify that any of the following functions are integrating factors:

(4.4) $$\frac{1}{xy}, \quad \frac{1}{x^2}, \quad \frac{1}{y^2}, \quad \frac{1}{x^2 + y^2},$$

except for values of x and y for which these fractions are meaningless. If we choose the first integrating factor, we have

$$\frac{dy}{y} - \frac{dx}{x} = 0,$$

and integral curves are given by

(4.5) $$\log |y| - \log |x| = \log |C|,$$

or,

$$\log \left| \frac{y}{x} \right| = \log |C|,$$

where C is a constant. The latter is equivalent to

$$y = Cx.$$

The method we employed breaks down when $xy = 0$; that is, along the coordinate axes. A reference to equation (4.3) shows that both $x = 0$ and $y = 0$ are also integral curves. We chose the constant in (4.5) in the form $\log |C|$ for convenience.

If we had chosen the integrating factor $\dfrac{1}{x^2}$, we would have been led to the equation

$$\frac{x \, dy - y \, dx}{x^2} = 0 \qquad (x \neq 0).$$

This may be rewritten as

$$d\left(\frac{y}{x} \right) = 0.$$

Integral curves are then readily seen to be given by the equation

$$\frac{y}{x} = C.$$

We must examine the curve $x = 0$ separately by reference to the original equation (4.3). Solving equation (4.3) by use of the remaining integrating factors in (4.4) is left to the reader.

Exercises

Solve the following differential equations.

1. $\sqrt{a^2 - y^2} \, dx + (b^2 + x^2) \, dy = 0.$

2. $(2x - 1)(y - 1)^2 \, dx + (x - x^2)(1 + y) \, dy = 0.$

3. $x \sin y \, dx + e^{-x} \, dy = 0.$

4. $\sin y \, dx + (1 - \cos x) \, dy = 0.$

5. $y \log x \, dx + (1 + 2y) \, dy = 0.$

6. $(2x - 1) \cos^4 y \, dx + (x^2 - 2x + 2) \, dy = 0.$

7. $y' = e^{x-y}.$

8. $e^{x+y} y' = e^{2x-y}.$

9. $e^{x^2-y^2} yy' = x + 2x^3.$

5   Homogeneous differential equations of first order

If the coefficients $M(x, y)$ and $N(x, y)$ of the differential equation

(5.1) $M(x, y) \, dx + N(x, y) \, dy = 0$ $[N(x, y) \neq 0]$

have the property that

(5.2) $\dfrac{M(ax, ay)}{N(ax, ay)} \equiv \dfrac{M(x, y)}{N(x, y)},$

where a is an arbitrary number not equal to zero, the equation (5.1) is said to have *homogeneous* coefficients. When this is true, either the substitution

(5.3) $y = vx, \qquad dy = v \, dx + x \, dv$

or

(5.3)′ $x = wy, \qquad dx = w \, dy + y \, dw$

will reduce (5.1) to an equation in which the variables are separable.

We shall verify this statement for the substitution (5.3). In that case (5.1) becomes

$$M(x, vx) \, dx + N(x, vx)(x \, dv + v \, dx) = 0.$$

This equation may be written

(5.4) $\dfrac{M(x, vx)}{N(x, vx)} \, dx + v \, dx + x \, dv = 0.$

Since equation (5.2) is an identity in the three variables a, x, and y, we may write

$$\frac{M(x, vx)}{N(x, vx)} \equiv \frac{M(1, v)}{N(1, v)}.$$

Accordingly, equation (5.4) may be rewritten in the form

$$[M(1, v) + vN(1, v)] \, dx + xN(1, v) \, dv = 0,$$

and the variables x and v are separable.

Example.   The differential equation

(5.1)′ $(x^2 + y^2)\, dx + 3xy\, dy = 0$

has homogeneous coefficients since

$$\frac{(ax)^2 + (ay)^2}{3(ax)(ay)} \equiv \frac{x^2 + y^2}{3xy} \qquad (axy \neq 0).$$

We may employ either substitution (5.3) or (5.3)′. If the former is used, equation (5.1)′ becomes

$$(x^2 + v^2 x^2)\, dx + 3x(vx)(v\, dx + x\, dv) = 0,$$

or

$$\frac{dx}{x} + \frac{3v\, dv}{1 + 4v^2} = 0.$$

The formal integration is easily managed, and we have

$$\log |x| + \tfrac{3}{8} \log (1 + 4v^2) = \tfrac{3}{8} \log c.$$

Setting $v = \dfrac{y}{x}$ we obtain

$$x^2(x^2 + 4y^2)^3 = c.$$

The substitution $x = wy$ would have led to a similar result.

An integrating factor.   When the equation

(5.5) $M(x, y)\, dx + N(x, y)\, dy = 0$

has homogeneous coefficients, an integrating factor is

$$\frac{1}{xM + yN}.$$

To verify this we require the following result, which is due to Euler.

Lemma. If the coefficients $M(x, y)$ and $N(x, y)$ of (5.5) are homogeneous and possess continuous partial derivatives in some region R of the xy-plane,

(5.6) $$\frac{xM_x + yM_y}{xN_x + yN_y} \equiv \frac{M}{N}.$$

To prove the lemma we differentiate both sides of the identity (5.2) partially with respect to a obtaining

(5.7) $N(ax, ay)[xM_x(ax, ay) + yM_y(ax, ay)]$
$$- M(ax, ay)[xN_x(ax, ay) + yN_y(ax, ay)] \equiv 0.$$

If we set $a = 1$ in (5.7), (5.6) follows at once.

The proof that $\dfrac{1}{xM + yN}$ is an integrating factor is now readily accomplished and will be left to the student as an exercise. Let us, however, apply this integrating factor to the last example. We have then

$$\frac{1}{xM + yN} = \frac{1}{x^3 + 4xy^2}$$

as an integrating factor for equation (5.1)'; that is, the equation

(5.8) $$\frac{x^2 + y^2}{x^3 + 4xy^2} \, dx + \frac{3xy}{x^3 + 4xy^2} \, dy = 0$$

is exact. We may use the *alternate* method of Section 2 and integrate formally the second term of (5.8), obtaining

$$\tfrac{3}{8} \log |x^3 + 4xy^2|.$$

Next, we seek a function $f(x)$, of x alone, such that

$$d(f(x) + \tfrac{3}{8} \log |x^3 + 4xy^2|)$$

is the left-hand member of (5.8). We see at once that

$$f'(x) = -\frac{1}{8x};$$

hence, $f(x)$ may be taken as $-\tfrac{1}{8} \log |x|$, and it follows that

$$x^2(x^2 + 4y^2)^3 = c.$$

Occasionally a differential equation with coefficients that are almost homogeneous can be rewritten in homogeneous form by use of simple transformations. For example, consider the differential equation

(5.9) $$(x - y - 1) \, dx + (x + 4y - 6) \, dy = 0.$$

The coefficients in this differential equation would be homogeneous if the constant terms -1 and -6 were not present. This situation can be remedied by setting

$$x = u + h, \qquad y = v + k,$$

where h and k are constants to be determined. Noting that $dx = du$ and $dy = dv$, we have

(5.10) $$[u - v + (h - k - 1)] \, du + [u + 2v + (h + 4k - 6)] \, dv = 0.$$

We attempt to choose h and k so that

$$h - k - 1 = 0,$$

$$h + 4k - 6 = 0.$$

When these simultaneous equations are solved, we find that $h = 2$, $k = 1$, and equation (5.10) becomes

(5.11) $$(u - v)\, du + (u + 4v)\, dv = 0,$$

which has homogeneous coefficients. This equation may be solved then by the substitution

$$u = wv, \qquad du = w\, dv + v\, dw.$$

After the usual computation, integral curves of equation (5.11) are found to be

$$\log (u^2 + 4v^2) = \arctan \left(\frac{u}{2v} \right) + C.$$

We now set $u = x - 2$, $v = y - 1$, and obtain

$$\log [(x - 2)^2 + 4(y - 1)^2] = \arctan \left[\frac{x - 2}{2(y - 1)} \right] + C$$

as an equation of integral curves of the given differential equation (5.9).

Exercises

Solve the following differential equations.

1. $(x - y)\, dx + x\, dy = 0$. (Use both methods of this section; also solve as a linear equation.)

2. $(x - 2y)\, dx + x\, dy = 0$. (Use both methods of this section.)

3. $(x^2 - y^2)\, dx + 2xy\, dy = 0$.

4. $\sqrt{x^2 + y^2}\, dx = x\, dy - y\, dx$.

5. $(x^2 y + 2xy^2 - y^3)\, dx - (2y^3 - xy^2 + x^3)\, dy = 0$.

6. $\left(x \sin \dfrac{y}{x} - y \cos \dfrac{y}{x} \right) dx + x \cos \dfrac{y}{x}\, dy = 0$.

7. $(x^3 + 2xy^2)\, dx + (y^3 + 2x^2 y)\, dy = 0$.

8. $(4x^4 - x^3 y + y^4)\, dx + x^4\, dy = 0$.

9. $\left(x^2 \sin \dfrac{y^2}{x^2} - 2y^2 \cos \dfrac{y^2}{x^2} \right) dx + 2xy \cos \dfrac{y^2}{x^2}\, dy = 0$.

10. $(x^2 e^{-y^2/x^2} - y^2)\, dx + xy\, dy = 0.$

11. $(2x + y - 2)\, dx + (2y - x + 1)\, dy = 0.$

12. $(x - 3y)\, dx + (x + y + 4)\, dy = 0.$

13. $(x + y - 10)\, dx + (x - y - 2)\, dy = 0.$

14. $(3y + x)\, dx + (x + 5y - 8)\, dy = 0.$

15. $(x - y)\, dx + (x - y + 2)\, dy = 0.$ (*Hint.* Set $u = x - y$.)

16. $(x + 2y + 1)\, dx + (2x + 4y + 3)\, dy = 0.$

17. $(x^2 + y^2)\, dx + kxy\, dy = 0$ (k constant).

18.   Solve the differential equation

$$(2x^3 + 4xy^3)\, dx + (6x^2 y^2 - 3y^5)\, dy = 0.$$

(*Hint.* Determine constants m, n so that the substitution $x = u^m$, $y = v^n$ reduces the equation to homogeneous form.)

Answers

1. $xe^{y/x} = c.$

3. $x^2 + y^2 = 2ax.$

5. $(x^2 - y^2)e^{x/y} = c.$

7. $x^4 + 4x^2 y^2 + y^4 = c.$

9. $x \sin (y^2/x^2) = c.$

11. $\log [(x - 1)^2 + y^2] - \arctan [y/(x - 1)] = c.$

13. $x^2 + 2xy - y^2 - 20x - 4y = c.$

15. $x + y - \log |x - y + 1| = c.$

17. $x[x^2 + (k + 1)y^2]^{k/2} = c \ (k \neq -1); \ y^2 = x^2 \log c^2 x^2 \ (k = -1).$

18. $2m - 3n = 0.$ For example, $m = 3, n = 2.$

6 Equations reducible to differential equations of first order

Consider the differential equation

(6.1) $y'' - y' = 0.$

This equation is of first order in the variable y'. This observation may be exploited by the use of the substitution

(6.2) $$p = y'.$$

Equation (6.1) then becomes the first-order differential equation

(6.1)' $$p' - p = 0$$

the solution of which is

$$p = c_1 e^x \qquad (c_1 \text{ constant}).$$

Thus, we set

$$y' = c_1 e^x,$$

and an integration yields

$$y = c_1 e^x + c_2 \qquad (c_2 \text{ constant}).$$

The substitution (6.2) will ordinarily be effective when the letter y is not present in the given differential equation. When the letter x is not present, a helpful substitution frequently is

$$y' = p,$$

(6.3)
$$y'' = \frac{dp}{dx} = \frac{dp}{dy}\frac{dy}{dx} = p\frac{dp}{dy}.$$

Example.   It has been noted that the solution of the differential equation

(6.4) $$y'' + y = 0$$

is $c_1 \sin x + c_2 \cos x$, where c_1 and c_2 are constants.

We observe that the variable x is not present formally in the equation and we apply the substitution (6.3). We have

$$p\frac{dp}{dy} + y = 0.$$

The variables are separable, and we write

$$p \, dp + y \, dy = 0;$$

hence

$$\frac{p^2}{2} + \frac{y^2}{2} = \frac{a_1^2}{2} \qquad (a_1 \text{ constant}, \neq 0).$$

Accordingly, we have

(6.5) $$y'^2 + y^2 = a_1^2,$$

and consequently,

$$\frac{dy}{dx} = \pm\sqrt{a_1^2 - y^2}.$$

Again the variables are separable, and we may write

$$\frac{dy}{\sqrt{a_1^2 - y^2}} = \pm\, dx.$$

An integration yields

$$\text{arc sin } \frac{y}{a_1} = \pm x + a_2,$$

or

(6.6) $$y = a_1 \sin\,(\pm x + a_2).$$

The equations (6.6) are readily seen to be equivalent to

(6.7) $$y = k_1 \sin\,(x - k_2),$$

where k_1 and k_2 are new constants. The solution of (6.4) provided by the
right-hand member of (6.7) may, if it is desired, be put in the form
$c_1 \sin x + c_2 \cos x$ since

$$k_1 \sin\,(x - k_2) = k_1(\sin x \cos k_2 - \cos x \sin k_2)$$
$$= k_1 \cos k_2 \sin x - k_1 \sin k_2 \cos x.$$

We set

$$c_1 = k_1 \cos k_2, \qquad c_2 = -k_1 \sin k_2.$$

It will be observed that the above method of treating equation (6.4) is
equivalent to multiplying both members of the equation by $2y'$:

$$2y'y'' + 2yy' = 0.$$

We then have

$$y'^2 + y^2 = a_1^2,$$

which is equation (6.5).

Exercises

Solve the following differential equations.

1. $y'' - 2y' = 0$ (two methods).

2. $y'' + y' = 0$ (two methods).

3. $xy'' - y' = 0$.

4. $xy'' + 3y' = 0$.

5. $y''' + y' = 0$.

6. $y''' - y' = 0$.

7. $y'' + a^2y = 0$.

8. $y'' = (1 + y'^2)^{3/2}$.

9. $2yy'' = 1 + y'^2, y(0) = 1, y'(0) = 0$.

Answers

1. $y = c_1 + c_2e^{2x}$.

3. $y = c_1 + c_2x^2$.

5. $y = c_1 + c_2 \sin x + c_3 \cos x$.

7. $y = c_1 \sin ax + c_2 \cos ax$.

9. $x^2 = 4(y - 1)$.

2

Linear differential equations with constant coefficients

1 Some generalities

In this chapter we shall be principally concerned with methods of solving linear differential equations having constant coefficients; that is, with equations of the form

(1.1) $$a_0 y^{(n)} + a_1 y^{(n-1)} + \cdots + a_{n-1} y' + a_n y = f(x),$$

where $a_0, a_1, \ldots, a_n$ are constants, and $f(x)$ is continuous on some interval. When $a_0 \neq 0$, equation (1.1) is of order n. With equation (1.1) we associate the corresponding *homogeneous* equation

(1.2) $$a_0 y^{(n)} + a_1 y^{(n-1)} + \cdots + a_{n-1} y' + a_n y = 0.$$

Solutions of (1.1) depend in a fundamental way on the solutions of (1.2), as we shall see. A *solution* of (1.1) is a function defined on an interval I with the property that when y is replaced by that function, equation (1.1) becomes an identity on I. The interval I may be open or closed at either end and may be either finite or infinite. A solution of (1.1) [and of (1.2)] necessarily possesses a continuous nth derivative on I.

Before attempting to solve equation (1.2), however, we shall consider the general *homogeneous linear differential equation*

(1.3) $$a_0(x) y^{(n)} + a_1(x) y^{(n-1)} + \cdots + a_{n-1}(x) y' + a_n(x) y = 0,$$

where $a_0(x)$, $a_1(x)$, ..., $a_n(x)$ are continuous and $a_0(x) \neq 0$ on an interval I.

Next, we introduce some concepts that are treated more carefully in Chapter 4. First, n solutions

$$(1.4) \qquad\qquad y_1(x), y_2(x), \ldots, y_n(x)$$

of (1.3) are said to be *linearly dependent* on the interval I if there exist n constants $c_1, c_2, \ldots, c_n$, not all zero, such that

$$c_1 y_1(x) + c_2 y_2(x) + \cdots + c_n y_n(x) \equiv 0$$

on I. If solutions (1.4) are not linearly dependent, they are said to be *linearly independent* on I. Thus, the functions

$$1, \qquad \sin x, \qquad 2 \sin x$$

are linearly dependent solutions of the equation

$$(1.5) \qquad\qquad y''' + y' = 0$$

on any interval of the x-axis, for

$$0 \cdot (1) + (-2) \sin x + 1 \cdot (2 \sin x) \equiv 0.$$

On the other hand, the solutions

$$1, \qquad \sin x, \qquad \cos x$$

of (1.5) form a linearly independent set of solutions. For, suppose there are constants c_1, c_2, c_3 such that

$$(1.6) \qquad\qquad c_1(1) + c_2 \sin x + c_3 \cos x \equiv 0$$

on an interval, which for simplicity we shall assume contains the interval $0 \leq x \leq \pi$. Since (1.6) is an identity, it must hold, in particular, when $x = 0$, $x = \pi/2$, and $x = \pi$; that is, we must have

$$c_1 + c_3 = 0,$$
$$c_1 + c_2 = 0,$$
$$c_1 - c_3 = 0.$$

It follows at once that $c_1 = c_2 = c_3 = 0$.

The following theorems are proved in Chapter 4.

Theorem 1.1.   A necessary and sufficient condition that n solutions $y_1(x)$, $y_2(x)$, ..., $y_n(x)$ of equation (1.3) be linearly dependent is that the determinant

$$\Delta(x) = \begin{vmatrix} y_1(x) & y_2(x) & \cdots & y_n(x) \\ y_1'(x) & y_2'(x) & \cdots & y_n'(x) \\ \cdot & \cdot & \cdot \cdots \cdot & \cdot \\ y_1^{(n-1)}(x) & y_2^{(n-1)}(x) & \cdots & y_n^{(n-1)}(x) \end{vmatrix} \equiv 0.$$

The determinant $\Delta(x)$ is called the *wronskian* of the given solutions.

*Theorem 1.2.   There exist n linearly independent solutions $y_1(x)$, $y_2(x)$, ...,
$y_n(x)$ of equation (1.3). Further, if $y_1(x)$, $y_2(x)$, ..., $y_n(x)$ are any n linearly
independent solutions of (1.3), every solution of (1.3) can be written in the form*

(1.7) $$c_1 y_1(x) + c_2 y_2(x) + \cdots + c_n y_n(x),$$

where $c_1, c_2, \ldots, c_n$ are constants.

The function (1.7) is called the *general solution* of equation (1.3), since
every solution can be written in this form, and every such linear combination
of solutions is a solution of (1.3). Clearly, the function 0 is always a solution
of (1.3). It is known as the *null solution*, or *trivial solution*.

If there is a value of x for which one or more of the functions $a_0(x)$, $a_1(x)$,
..., $a_n(x)$ fails to be continuous or for which $a_0(x)$ vanishes, it is called a
singular point of equation (1.3).

We return to equation (1.2), which we repeat for easy reference:

(1.2) $$a_0 y^{(n)} + a_1 y^{(n-1)} + \cdots + a_{n-1} y' + a_1 y = 0 \qquad (a_0 \neq 0).$$

The numbers $a_0, a_1, \ldots, a_n$ are constants. A special case is the first-order
equation

(1.8) $$y' + a_1 y = 0,$$

a solution of which is, as we have seen, $e^{-a_1 x}$. Equation (1.8) could have been
solved by attempting to find a constant m such that e^{mx} is a solution of (1.8).
Substituting e^{mx} for y in (1.8) leads to the equation

$$e^{mx}(m + a_1) = 0$$

for the determination of m. Inasmuch as e^{mx} is never zero, we observe that
$m = -a$, and we have the solution $e^{-a_1 x}$ previously noted.

The foregoing suggests that we attempt to find a solution of (1.2) by
substituting

$$y = e^{mx}$$

in that equation. When this is done, we are led to the equation

$$e^{mx}(a_0 m^n + a_1 m^{n-1} + \cdots + a_{n-1} m + a_n) = 0,$$

or

(1.9) $$a_0 m^n + a_1 m^{n-1} + \cdots + a_{n-1} m + a_n = 0,$$

for the determination of m. Clearly, if m_1 is any real root of equation (1.9) [called the *auxiliary* or *indicial* or *characteristic* equation], the function $e^{m_1 x}$ is a solution of (1.2).

Equation (1.9) is a polynomial equation in m of degree n. It will thus have n roots. The treatment of equation (1.2) varies according to whether the roots of the indicial equation are real or imaginary and also according to whether the equation possesses or does not possess multiple roots.

2   Real roots

If the roots of (1.9) are the n real, distinct numbers

$$m_1, m_2, \ldots, m_n,$$

the functions

(2.1) $$e^{m_1 x}, e^{m_2 x}, \ldots, e^{m_n x}$$

are a set of n solutions for which $\Delta(x) \neq 0$, and the general solution of (1.2) is

(2.2) $$c_1 e^{m_1 x} + c_2 e^{m_2 x} + \cdots + c_n e^{m_n x}.$$

The proof that the solutions (2.1) are such that $\Delta(x) \neq 0$ is suggested by Exercise 22 of Section 3.

Example.   Find the general solution of

(2.3) $$y'' + y' - 2y = 0.$$

The auxiliary equation

(1.4) $$m^2 + m - 2 = 0$$

has the roots $1, -2$, which are real and distinct. The general solution of (2.3) is then

$$y = c_1 e^x + c_2 e^{-2x}.$$

Consider the differential equation

$$(2.4) \qquad y'' - 2y' + y = 0.$$

The auxiliary equation

$$(2.5) \qquad m^2 - 2m + 1 = 0$$

has the roots 1, 1. It is clear that e^x is then a solution of (2.4). A second solution needs to be found—a solution with the property that the wronskian of it and e^x is not identically zero. To this end, we attempt to find a function $v(x)$ such that $e^x v(x)$ is a solution of equation (2.1). The substitution $y = e^x v$ in (2.4) yields

$$(2.6) \qquad e^x v'' = 0.$$

Inasmuch as e^x is never zero, this equation reduces to the equation

$$(2.7) \qquad v'' = 0.$$

A solution of this differential equation is the function x, and the corresponding solution $e^x v$ of (2.4) is

$$xe^x.$$

The wronskian of the pair of solutions x and xe^x is

$$\begin{vmatrix} e^x & xe^x \\ e^x & e^x(1 + x) \end{vmatrix} = e^{2x} \neq 0,$$

and the general solution of (2.4) is, accordingly,

$$(2.8) \qquad c_1 e^x + c_2 x e^x.$$

It should be noted that the general solution of (2.7) is $ax + b$, where a and b are constants. The product

$$e^x v = e^x(ax + b),$$

however, does not (indeed, *cannot*) yield anything more than (2.8).

We anticipate the following generalization. If, for example, the differential equation is of third order and leads to an auxiliary equation having 2, 2, 2 as roots, the general solution of the differential equation is

$$y = c_1 e^{2x} + c_2 x e^{2x} + c_3 x^2 e^{2x},$$

and so on. The student can easily construct the differential equation which would have led to this solution.

Finally, if, for example, the roots of the auxiliary equation were the set

$$2, \ 2, \ 2, \ -3, \ -3, \ -\tfrac{1}{2},$$

the general solution of the corresponding differential equation would be

$$y = c_1 e^{2x} + c_2 x e^{2x} + c_3 x^2 e^{2x} + c_4 e^{-3x} + c_5 x e^{-3x} + c_6 e^{-x/2}.$$

3   Imaginary roots

If the auxiliary equation has the imaginary root $a + ib$ (a, b real, $b \neq 0$), it also has the *conjugate* root $a - ib$. Corresponding linearly independent solutions of the differential equation are $e^{ax} \cos bx$ and $e^{ax} \sin bx$.

In order to understand how these solutions arise we borrow from the theory of functions of a complex variable. First, it is shown there that

$$\frac{d}{dx} e^{cx} = c e^{cx},$$

even when c is an imaginary number. Accordingly, if $c = a + ib$ is a root of the auxiliary algebraic equation, the function e^{cx} formally satisfies the differential equation, and so does e^{hx}, where $h = a - ib$. Also, the sum and difference of these functions satisfy the differential equation; that is,

$$e^{cx} + e^{hx} \quad \text{and} \quad e^{cx} - e^{hx}$$

formally satisfy the differential equation.

Next, we borrow from complex analysis the Euler identity which states that

$$e^{(a+ib)x} = e^{ax} e^{ibx} = e^{ax}(\cos bx + i \sin bx),$$

$$e^{(a-ib)x} = e^{ax} e^{-ibx} = e^{ax}(\cos bx - i \sin bx)$$

for all values of x. We have then that the functions

$$e^{cx} + e^{hx} = 2 e^{ax} \cos bx,$$

$$e^{cx} - e^{hx} = 2i e^{ax} \sin bx$$

formally satisfy the differential equation, and hence that the functions

$$e^{ax} \cos bx \quad \text{and} \quad e^{ax} \sin bx$$

satisfy the differential equation.

Example.   Consider the differential equation

(3.1)
$$y'' - 4y' + 13y = 0.$$

The corresponding auxiliary equation has roots $c = 2 + 3i$ and $h = 2 - 3i$.

According to our theory, the functions

$$e^{2x} \sin 3x \quad \text{and} \quad e^{2x} \cos 3x$$

should be solutions of (3.1). Let us verify that the former function is indeed a solution. We have

$$y = e^{2x} \sin 3x,$$

$$y' = e^{2x}(3 \cos 3x + 2 \sin 3x),$$

$$y'' = e^{2x}(12 \cos 3x - 5 \sin 3x).$$

It is readily seen that $e^{2x} \sin 3x$ is a solution of (3.1). A similar computation will show that $e^{2x} \cos 3x$ is also a solution.

If the auxiliary equation has the repeated imaginary roots

$$a + ib, \, a + ib, \, a - ib, \, a - ib,$$

it can be verified that the general solution of the differential equation is

$$c_1 e^{ax} \cos bx + c_2 e^{ax} \sin bx + c_3 x e^{ax} \cos bx + c_4 x e^{ax} \sin bx.$$

Example. Consider the differential equation

(3.2) $$y'''' - 8y''' + 42y'' - 104y' + 169y = 0.$$

Its auxiliary equation

$$m^4 - 8m^3 + 42m^2 - 104m + 169 = 0$$

has the roots $2 + 3i, \, 2 + 3i, \, 2 - 3i, \, 2 - 3i$. The general solution of (3.2) is then

$$c_1 e^{2x} \cos 3x + c_2 e^{2x} \sin 3x + c_3 x e^{2x} \cos 3x + c_4 x e^{2x} \sin 3x.$$

Exercises

Prove that the given functions are linearly independent solutions of the given differential equation by using directly the definition of linear independence and by computing the wronskian.

1. $y'' + y = 0,$ $\sin x, \cos x$ $(-\infty < x < \infty).$

2. $y'' + y = 0,$ $\sin x, \sin x - \cos x$ $(-\infty < x < \infty).$

3. $y'' - y = 0,$ e^x, e^{-x} $(-\infty < x < \infty).$
   [*Hint.* Differentiate the identity $c_1 y_1(x) + c_2 y_2(x) \equiv 0$].

4. $y'' - 2y' + y = 0,$ e^x, xe^x $(0 \leq x \leq 3).$

5. $y''' + y' = 0,$ $-2, \sin x, 3 \cos x$ $(-\pi \leq x \leq \pi).$

6. $y'''' - y = 0,$ $e^x, e^{-x}, \sin x, \cos x$ $(-\infty < x < \infty).$

Find the general solutions of the following differential equations.†

7. $y'' - 3y' + 2y = 0.$

8. $y'' + 2y' + 2y = 0.$

9. $y'' + 4y' + 4y = 0.$

10. $y'' + 4y' + 8y = 0.$

11. $y'' + 6y' + 9y = 0.$

12. $y'' - 4y' + 3y = 0.$

13. $8y'' + 4y' + y = 0.$

14. $4y'' + 4y' + y = 0.$

15. $y''' - 3y'' + 4y' - 2y = 0.$

16. $y''' - y'' - y' + y = 0.$

17. $y'''' + 4y = 0.$

18. $y'''' - 6y''' + 19y'' - 26y' + 18y = 0.$

19.   A homogeneous linear differential equation with constant coefficients has

$$2, 2, 2, 3 - 4i, 3 + 4i, 3 - 4i, 3 + 4i, 3, 3$$

as roots of its auxiliary equation. What is its general solution?

20.   Do the same as Exercise 19 when the roots are

$$2, 2, 3 - 4i, 3 + 4i, 3 - 4i, 3 + 4i, 3 - 4i, 3 + 4i, 7.$$

21.   Show that the wronskian of $e^{ax} \cos bx$ and $e^{ax} \sin bx$ $(b \neq 0)$ is be^{2ax} (or $-be^{2ax}$).

22.   Prove that the wronskian of the functions

$$e^{m_1 x}, e^{m_2 x}, e^{m_3 x}$$

is equal to $(m_1 - m_2)(m_2 - m_3)(m_3 - m_1)e^{(m_1 + m_2 + m_3)x}$. From this result guess the value of the wronskian of the n functions (2.1).

23.   The indicial equation associated with the differential equation

$$y'' - 2ay' + (a^2 + b^2)y = 0 \qquad (b \neq 0)$$

has roots $a \pm ib$. Show, by substitution, that $e^{ax} \sin bx$ and $e^{ax} \cos bx$ are solutions of the differential equation. Use Theorem 1.1 to show that these solutions are linearly independent.

---

† Exercises 17 and 18 require some ingenuity in solving the associated auxiliary equation.

Answers

7. $c_1 e^x + c_2 e^{2x}$.

8. $e^{-x}(c_1 \cos x + c_2 \sin x)$.

9. $c_1 e^{-2x} + c_2 x e^{-2x}$.

10. $e^{-2x}(c_1 \cos 2x + c_2 \sin 2x)$.

11. $c_1 e^{-3x} + c_2 x e^{-3x}$.

12. $c_1 e^x + c_2 e^{3x}$.

13. $e^{-x/4}(c_1 \cos \frac{1}{4}x + c_2 \sin \frac{1}{4}x)$.

14. $c_1 e^{-x/2} + c_2 x e^{-x/2}$.

15. $e^x(c_1 + c_2 \cos x + c_3 \sin x)$.

16. $c_1 e^{-x} + c_2 e^x + c_3 x e^x$.

17. $e^x(c_1 \cos x + c_2 \sin x) + e^{-x}(c_3 \cos x + c_4 \sin x)$.

18. $e^x(c_1 \cos x + c_2 \sin x) + e^{2x}(c_3 \cos (x\sqrt{5}) + c_4 \sin (x\sqrt{5}))$.

4   The nonhomogeneous case

Consider the differential equation

(4.1) $$y'' - 3y' + 2y = 4.$$

With (4.1) we associate the homogeneous equation

(4.2) $$y'' - 3y' + 2y = 0.$$

The general solution of (4.2) is

$$c_1 e^x + c_2 e^{2x}.$$

We wish to find the general solution of (4.1). To obtain it we make use of the following result, which is proved in Chapter 4.

Theorem 4.1. If $c_1 y_1(x) + c_2 y_2(x)$ is the general solution of

$$a(x)y'' + b(x)y' + c(x)y = 0,$$

and if $y_0(x)$ is any particular solution of the nonhomogeneous differential equation

(4.3) $$a(x)y'' + b(x)y' + c(x)y = f(x),$$

the general solution of (4.3) *is*

$$y_0(x) + c_1 y_1(x) + c_2 y_2(x).$$

We assume that $a(x)$, $b(x)$, $c(x)$, $f(x)$ are continuous on some interval I, with $a(x) \neq 0$ on I.

Theorem 4.1 is stated for second-order linear differential equations. A similar statement is valid for linear differential equations of order n ($n = 1, 2, \ldots$).

We return to equation (4.1). By inspection, we see that a particular solution of (4.1) is $y = 2$. Accordingly, the general solution of (4.1) is

$$2 + c_1 e^x + c_2 e^{2x}.$$

When a particular solution of a nonhomogeneous differential equation cannot be found by inspection, other methods are available. A general method, called variation of parameters, is discussed in Chapter 4. In the present chapter we confine our attention to certain special devices that are sometimes useful. Roughly speaking, we can apply these devices when the right-hand member of the equation is composed of the sum of terms each of which possesses a finite number of essentially different derivatives. For example, the right-hand member of the nonhomogeneous differential equation

$$(4.4) \qquad\qquad y'' - y' - 2y = -2e^x - 10 \cos x$$

contains the terms e^x and $\cos x$ (except for constant multipliers). Except for constant factors, these terms and their possible derivatives are e^x, $\sin x$, $\cos x$. The plan is to try to determine constants a, b, and c so that the function

$$ae^x + b \sin x + c \cos x$$

is a solution of (4.4). If this function is substituted for y in (4.4), we have

$$-2ae^x + (c - 3b) \sin x - (3c + b) \cos x = -2e^x - 10 \cos x.$$

We then have the equations†

$$-2a = -2,$$
$$c - 3b = 0,$$
$$3c + b = 10$$

---

† Although it is true that the following equations must hold if the preceding equation is to be an identity, we need not be concerned with this. The problem is clearly solved if any constants a, b, and c can be found such that $ae^x + b \sin x + c \cos x$ is a solution of (4.4).

to determine a, b, and c. It follows that $a = 1$, $b = 1$, $c = 3$, and a solution of (4.4) is

$$e^x + \sin x + 3 \cos x.$$

Since the general solution of the corresponding homogeneous equation

$$y'' - y' - 2y = 0$$

is

$$c_1 e^{-x} + c_2 e^{2x},$$

the general solution of (4.4) is

$$e^x + \sin x + 3 \cos x + c_1 e^{-x} + c_2 e^{2x}.$$

Example. Solve the differential equation

(4.5) $$\qquad\qquad y'' + y = 3.$$

The general solution of the corresponding homogeneous equation

$$y'' + y = 0$$

is $c_1 \sin x + c_2 \cos x$. By inspection we see that a particular solution of (4.5) is 3; thus, the general solution of (4.5) is

$$3 + c_1 \sin x + c_2 \cos x.$$

Example. Solve the differential equation

(4.6) $$\qquad\qquad y'' + 4y = -4x^2 + 2.$$

The general solution of the corresponding homogeneous equation is seen to be $c_1 \sin 2x + c_2 \cos 2x$. We shall try for a solution of (4.6) of the form $y = ax^2 + bx + c$. We have

$$4ax^2 + 4bx + (2a + 4c) = -4x^2 + 2.$$

Thus, we try to determine constants a, b, and c such that

$$4a = -4,$$

$$4b = 0,$$

$$2a + 4c = 2.$$

It follows that we may set $a = -1$, $b = 0$, $c = 1$, and a particular solution of (4.6) is then $-x^2 + 1$. Its general solution is

$$-x^2 + 1 + c_1 \sin 2x + c_2 \cos 2x.$$

Example. Consider the differential equation

(4.7) $$\qquad\qquad y'' - y = 2e^x.$$

The general solution of the corresponding homogeneous equation

(4.8) $y'' - y = 0$

is

$$c_1 e^x + c_2 e^{-x}.$$

We try to determine the constant $a \neq 0$ so that $y = ae^x$ is a solution of (4.7). Upon the substitution in (4.7) we have the equation

$$0 = 2e^x$$

for the determination of the constant a. This is not possible, and another method of solution must be attempted.

The method we have been employing failed for equation (4.7) because the term e^x in the right-hand member is also a solution of the homogeneous equation (4.8). The difficulty can be circumvented by the device of determining a constant $a \neq 0$ such that $y = axe^x$ is a solution of (4.7). We have then

$$2ae^x = 2e^x.$$

It follows that $a = 1$ and that a particular solution of (4.7) is xe^x. Its general solution is, then,

$$xe^x + c_1 e^x + c_2 e^{-x}.$$

Example.   Solve the differential equation

(4.9) $y'' + y = 2e^x + 4 \sin x.$

The general solution of the corresponding homogeneous equation is $c_1 \sin x + c_2 \cos x$. We observe that the term $4 \sin x$ in the right-hand member of (4.9) is a solution of the corresponding homogeneous equation. This suggests that we try for a particular solution of (4.9) of the form

(4.10) $y = ae^x + bx \sin x + cx \cos x,$

where a, b, and c are constants. Upon substituting (4.10) in (4.9) we have

$$2ae^x + 2b \cos x - 2c \sin x = 2e^x + 4 \sin x,$$

and we may take

$$a = 1, \qquad b = 0, \qquad c = -2.$$

The general solution of (4.9) is, accordingly,

$$e^x - 2x \cos x + c_1 \sin x + c_2 \cos x.$$

If the given equation were

$$y'' + y = x \sin x + 2e^{2x},$$

we should try for a particular solution of the form

$$y = c_1 e^{2x} + c_2 x \sin x + c_3 x \cos x.$$

The method above would fail, for example, if the equation were

$$y'' + y = \tan x,$$

since $\tan x$ does not have only a finite number of essentially different forms for its derivatives.

Composition of solutions.   To find a particular solution of the differential equation

(4.11) $$y'' + 4y = e^x - 2 \sin x + x,$$

we may determine constants c_1, c_2, c_3, c_4, c_5 such that the function

$$c_1 e^x + c_2 \sin x + c_3 \cos x + c_4 x + c_5$$

will be a solution of (4.11). We may also find particular solutions of each of the following differential equations:

$$y'' + 4y = e^x,$$
$$y'' + 4y = -2 \sin x,$$
$$y'' + 4y = x.$$

The sum of these three particular solutions will be seen to be a solution of (4.11). The validity of this method is readily established. The proof for the case of a second-order linear differential equation is left to the student as an exercise (Exercise 17).

Exercises

Find the general solutions of the following differential equations.

1. $y'' - 3y' + 2y = \sin x.$

2. $y'' + 2y' + 2y = 1 + x^2.$

3. $y'' + 4y' + 4y = x - 2e^{2x}.$

4. $y'' + 4y' + 8y = x - e^x.$

5. $y'' - 6y' + 9y = e^x \sin x.$

6. $y'' - 4y' + 3y = x^3.$

7. $8y'' + 4y' + y = \sin x - 2 \cos x.$

8. $4y'' + 4y' + y = e^x - 2 \cos 2x$.

9. $y'' + y = 3 \cos x$.

10. $y'' + y = e^x + 3 \cos x$.

11. $y'' - 3y' + 2y = 2 + e^x$.

12. $y'' - 3y' + 2y = e^x - 2e^{2x} + \sin x$.

13. $y'' - 9y = 18x + 6e^{3x}$.

14. $y''' + 4y = x^2 - 3x + 2$.

15. $y^{(6)} - y = 2 \sin x - 4 \cos x$.

16. $y'' - 2y' + 2y = 2e^x \sin x$.

17. Establish the validity of the method of addition of solutions for the equation
$$y'' + a(x)y' + b(x)y = f_1(x) + f_2(x).$$

18. Find the general solution of
$$y'' - 2y' + y = 4e^x.$$

19. Find a particular solution of the differential equation (4.7) by means of the substitution $y = e^x v$.

20. Use Exercise 17 and the method of Exercise 19 to solve the differential equation (4.9).

Answers

1. $\frac{1}{10} \sin x + \frac{3}{10} \cos x + c_1 e^x + c_2 e^{2x}$.

2. $1 - x + \frac{1}{2}x^2 + e^{-x}(c_1 \cos x + c_2 \sin x)$.

3. $-\frac{1}{4} + \frac{1}{4}x - \frac{1}{2}e^{2x} + c_1 e^{-2x} + c_2 x e^{-2x}$.

4. $-\frac{1}{16} + \frac{1}{8}x - \frac{1}{13}e^x + e^{-2x}(c_1 \cos 2x + c_2 \sin 2x)$.

5. $e^x(\frac{3}{25} \sin x + \frac{4}{25} \cos x) + c_1 e^{3x} + c_2 x e^{3x}$.

6. $\frac{80}{27} + \frac{16}{9}x + \frac{4}{3}x^2 + \frac{1}{3}x^3 + c_1 e^x + c_2 e^{3x}$.

7. $\frac{3}{13} \cos x - \frac{2}{13} \sin x + e^{-x/4}(c_1 \cos \frac{1}{4}x + c_2 \sin \frac{1}{4}x)$.

8. $\frac{1}{9}e^x + \frac{30}{189} \cos 2x - \frac{16}{189} \sin 2x + c_1 e^{-x/2} + c_2 x e^{-x/2}$.

9. $\frac{3}{4}x \sin x + c_1 \cos x + c_2 \sin x$.

10. $\frac{1}{2}e^x + \frac{3}{2}x \sin x + c_1 \cos x + c_2 \sin x$.

11. $1 - xe^x + c_1 e^x + c_2 e^{2x}.$

12. $\frac{1}{10} \sin x + \frac{3}{10} \cos x - xe^x - 2xe^{2x} + c_1 e^x + c_2 e^{2x}.$

13. $xe^{3x} - 2x + c_1 e^{3x} + c_2 e^{-3x}.$

15. $-\sin x + 2 \cos x + c_1 e^x + c_2 e^{-x}$

$$+ e^{-x/2}\left(c_3 \cos \frac{\sqrt{3}x}{2} + c_4 \sin \frac{\sqrt{3}x}{2}\right) + e^{x/2}\left(c_5 \cos \frac{\sqrt{3}x}{2} + c_6 \sin \frac{\sqrt{3}x}{2}\right).$$

18. $2x^2 e^x + c_1 e^x + c_2 xe^x.$

5 Euler-type equations

An equation of the form

$$(5.1) \quad a_0 x^n y^{(n)} + a_1 x^{n-1} y^{(n-1)} + \cdots + a_{n-1} xy' + a_n y = f(x) \qquad (a_0 \neq 0),$$

where $a_0, a_1, \ldots, a_n$ are constants and $f(x)$ is continuous on some interval I not containing the origin, is known as an Euler-type equation. The corresponding homogeneous equation is

$$(5.2) \qquad a_0 x^n y^{(n)} + a_1 x^{n-1} y^{(n-1)} + \cdots + a_{n-1} xy' + a_n y = 0.$$

Inasmuch as the interval I does not contain the origin, x is either always positive or always negative on I. In the former case, the substitution $x = e^t$ reduces (5.2) to a linear differential equation with constant coefficients, while in the latter case the substitution $x = -e^t$ has the same effect. In both cases, we have (see Exercise 18)

$$(5.3) \qquad xy' = \dot{y}, \ x^2 y'' = \ddot{y} - \dot{y}, \ x^3 y''' = \dddot{y} - 3\ddot{y} + 2\dot{y}, \ldots.$$

Consider the differential equation

$$(5.4) \qquad\qquad 2x^2 y'' + 3xy' - y = 0.$$

Whether x is positive or negative on I, the appropriate transformation leads to the differential equation

$$(5.5) \qquad\qquad 2\ddot{y} + \dot{y} - y = 0,$$

with the general solution

$$(5.6) \qquad\qquad c_1 e^{-t} + c_2 e^{t/2}.$$

Inasmuch as both substitutions can be written as

$$|x| = e^t \quad \text{or} \quad t = \log |x|,$$

the solution (5.6) then becomes

(5.7) $$c_1 |x|^{-1} + c_2 |x|^{1/2},$$

whether x is always positive or always negative on I. When $x > 0$ on I, the general solution (5.7) may be rewritten as

$$c_1 x + c_2 x^{1/2},$$

while if $x < 0$ on I, it becomes

$$c_1 x + c_2 (-x)^{1/2}.$$

Consider next the nonhomogeneous equation

(5.8) $$x^2 y'' - 5xy' + 25y = 25 \log (-x).$$

It is clear that we are necessarily concerned with $x < 0$. The substitution

(5.9) $$x = -e^t, \quad t = \log (-x)$$

yields, with the help of (5.3), the differential equation

(5.10) $$\ddot{y} - 6\dot{y} + 25y = 25t.$$

The corresponding homogeneous equation is

(5.10)' $$\ddot{y} - 6\dot{y} + 25y = 0,$$

with the general solution

$$c_1 e^{3t} \cos 4t + c_2 e^{3t} \sin 4t.$$

A particular solution of (5.10) is readily found to be

$$t + \tfrac{6}{25};$$

accordingly, the general solution of (5.10) is

$$\tfrac{6}{25} + t + e^{3t}(c_1 \cos 4t + c_2 \sin 4t).$$

Referring to (5.3) we then have

$$\tfrac{6}{25} + \log (-x) + x^3 \{ c_1 \cos [4 \log (-x)] + c_2 \sin [4 \log (-x)] \}$$

as the general solution of (5.8).

An alternate method. The form of the solution of (5.4), for example, suggests that one should attempt to solve such an equation directly by trying to determine a constant m so that x^m is a solution of (5.2), or of (5.4). We then have the equation

$$x^m[2m^2 + m - 1] = 0,$$

or

(5.11) $$2m^2 + m - 1 = 0$$

for the determination of m. This last equation is called the *indicial* (or *characteristic*) *equation* associated with (5.4). Its (indicial, or characteristic) roots are $m = \frac{1}{2}, -1$, and the general solution, for $x > 0$, is

$$c_1 x^{-1} + c_2 x^{1/2}.$$

The substitution $(-x)^m$ or $|x|^m$ in (5.4) would have led to the same indicial equation (5.11), so the general solution of (5.4) is seen to be

$$c_1 x^{-1} + c_2 |x|^{1/2},$$

whether $x > 0$ or $x < 0$, on I.

This alternate method is usually simpler than the earlier method when the indicial roots are real and distinct. It is also available when there are imaginary or repeated indicial roots, but this requires additional analysis and a bit more of the theory of functions of a complex variable. As a practical procedure, the best plan is probably to begin by substituting $y = x^m$ in the (homogeneous) differential equation and obtaining the indicial equation. If its roots are real and distinct, solve the differential equation directly; otherwise, make the substitution $|x| = e^t$.

Exercises

Find the general solution of the given differential equations.

1. $x^2 y'' - 3xy' + 3y = 0.$

2. $x^2 y'' - xy' + 2y = 0.$

3. $x^2 y'' + 3xy' + 2y = 0.$

4. $x^2 y'' - 2y = 0.$

5. $x^2y'' + 5xy' + 8y = 0.$

6. $x^2y'' - 3xy' + 3y = \log |x|.$

7. $x^2y'' - xy' + 2y = 1 + \log^2 |x|.$

8. $x^2y'' - 2xy' + 2y = \sin \log |x|.$

9. $x^3y''' - 3x^2y'' + 6xy' - 6y = 0.$

10. $4x^2y'' - 4xy' + 3y = 0.$

11. $4x^2y'' - 4xy' + 3y = \sin \log (-x)$ $(x < 0).$

12.   Solve the differential equation $x^2y'' + a^2y = 0.$ (Identify the three cases: $a^2 < \frac{1}{4}, a^2 = \frac{1}{4}, a^2 > \frac{1}{4}.$)

13.   Solve the differential equation

$$x^3y''' - 2x^2y'' + 13xy' - 13y = 0   (x > 0).$$

14.   Solve the differential equation

$$x^3y''' - 2x^2y'' + 13xy' - 13y = 9x^2 + 26   (x > 0).$$

15.   Show that the wronskian of the functions

$$x^a, x^b, x^c   (x > 0)$$

is equal to $(a - b)(b - c)(c - a)x^{a+b+c-3}$. (Hence, the wronskian is never zero when the numbers a, b, and c are distinct.)

16.   Find the wronskian of the functions

$$x^m \sin \log x^n, x^m \cos \log x^n   (x > 0; m, n \text{ constants}).$$

17.   Determine an equation of the form $x^2y'' + bxy' + cy = 0$ that has the functions in Exercise 16 as solutions.

18.   Verify equations (5.3).

Answers

1. $c_1x + c_2x^3.$

2. $x(c_1 \cos \log |x| + c_2 \sin \log |x|).$

3. $x^{-1}(c_1 \cos \log |x| + c_2 \sin \log |x|).$

4. $c_1x^{-1} + c_2x^2.$

5. $x^{-2}(c_1 \cos \log x^2 + c_2 \sin \log x^2)$.

6. $\frac{4}{9} + \frac{1}{3} \log |x| + c_1 x + c_2 x^3$.

7. $1 + \log |x| + \frac{1}{2} \log^2 |x| + x(c_1 \cos \log |x| + c_2 \sin \log |x|)$.

8. $\frac{1}{10} \sin \log |x| + \frac{3}{10} \cos \log |x| + c_1 x + c_2 x^2$.

9. $c_1 x + c_2 x^2 + c_3 x^3$.

10. $c_1 |x|^{1/2} + c_2 |x|^{3/2}$.

11. $-\frac{1}{65} \sin \log (-x) + \frac{8}{65} \cos \log (-x) + c_1 (-x)^{1/2} + c_2 (-x)^{3/2}$.

13. $c_1 x + c_2 x^2 \sin \log x^3 + c_3 x^2 \cos \log x^3$.

17. $x^2 y'' + (1 - 2m)xy' + (m^2 + n^2)y = 0$.

3

The fundamental existence theorem

1 Introduction

In this chapter we shall first be concerned with the existence and behavior of solutions of differential equations of the form

$$(1.1) \qquad\qquad y^{(n)} = f(x, y, y', \ldots, y^{(n-1)}),$$

where $y^{(k)}$ indicates the kth derivative $d^k y/dx^k$, and f is a function of its $n+1$ arguments defined in a domain† D. By a *solution $y(x)$* of (1.1) will be meant a function of class C^n *defined on an interval* $I \subset D$ of the x-axis with the property that the substitution of $y(x)$ for y in (1.1) reduces that equation to an identity on I. The interval I, which is finite or infinite, may be any one of the types $[a, b]$, (a, b), $[a, b)$, or $(a, b]$. A function is said to be of *class C^n* on I if it and its first n derivatives are continuous on I.

Thus,

$$y' = y$$

is a differential equation of first order that has the solution ce^x on every interval for each fixed value of the constant c. Similarly, a solution of the differential equation

$$y' = e^{x^2}$$

is seen to be

---

† The domain D is an open connected set in R_{n+1}.

$$c + \int_0^x e^{x^2}\, dx,$$

for each fixed value of the constant c.

The second-order differential equation

$$y'' = -y$$

is readily seen to have solutions $\sin x$ and $\cos x$. Indeed, it has the solution

$$a \sin x + b \cos x$$

on every interval of the x-axis, for every choice of the pair of constants a and b. One notes that the choice $a = b = 0$ yields the *null solution*, that is, the function 0.

If the substitution

(1.2)
$$y = y_1,$$
$$y' = y_2,$$
$$\cdot \quad \cdot \quad \cdot \quad \cdot$$
$$y^{(n-1)} = y_n$$

is made in (1.1), that equation is transformed into the system

(1.3)
$$\frac{dy_1}{dx} = y_2,$$

$$\frac{dy_2}{dx} = y_3,$$

$$\cdot \quad \cdot \quad \cdot \quad \cdot \quad \cdot \quad \cdot \quad \cdot$$

$$\frac{dy_{n-1}}{dx} = y_n,$$

$$\frac{dy_n}{dx} = f(x, y_1, y_2, \ldots, y_n)$$

of n simultaneous first-order differential equations.

The system (1.3) may, of course, also be written as

$$y_1' = y_2,$$
$$y_2' = y_3,$$
$$\cdot \quad \cdot \quad \cdot \quad \cdot \quad \cdot \quad \cdot \quad \cdot$$
$$y_{n-1}' = y_n,$$
$$y_n' = f(x, y_1, y_2, \ldots, y_n).$$

This system is clearly a special case of the more general system

$$
\begin{aligned}
y_1' &= f_1(x, y_1, y_2, \ldots, y_n),\\
y_2' &= f_2(x, y_1, y_2, \ldots, y_n),\\
&\cdots \cdots \cdots \cdots\\
y_n' &= f_n(x, y_1, y_2, \ldots, y_n),
\end{aligned}
$$

(1.4)

where the functions f_i are real functions of their $n + 1$ arguments in a common domain D. For many purposes the system (1.4) can be written conveniently in shorthand notation as

(1.5) $$y' = f(x, y),$$

where y is understood to mean the vector $(y_1, y_2, \ldots, y_n)$, and $f(x, y)$ is the vector $(f_1, f_2, \ldots, f_n)$.

By a *solution* of (1.4) is meant a column of functions

$$
\begin{aligned}
y_1(x)\\
y_2(x)\\
\cdots\\
y_n(x)
\end{aligned}
$$

each of class C^n on a common interval I of the x-axis, the substitution of which in (1.4) reduces each equation to an identity in I. A solution of the *vector differential equation* (1.5) is then a vector $y(x)$ of class C^n on I that satisfies the differential equation (1.5). The language has changed slightly, but the meaning is the same.

The system

(1.6)

$$
\frac{dy_1}{dx} = 3y_1 - y_2,
$$

$$
\frac{dy_2}{dx} = 2y_1,
$$

for example, is readily seen to have as a solution the column of functions

$$
\begin{aligned}
e^x\\
2e^x.
\end{aligned}
$$

That is, if in (1.6) y_1 is replaced by e^x and y_2 by $2e^x$, equations (1.6) become identities on every interval of the x-axis.

Although we shall on occasion solve some differential equations, that is not the primary concern of the book. Nor shall we be concerned other than parenthetically with the fact that the theory of ordinary differential equations has many important applications in such areas as mechanics, engineering, biology, and economics. Our principal pursuit will be the behavior of solutions of single differential equations and of systems of differential equations without solving such equations. For example, without knowing any solution of the equation

$$y'' + xy = 0,$$

we shall be able to show quite simply that every solution of this equation must vanish infinitely often on the interval $(-\infty, \infty)$.

2   The fundamental existence theorem

It is easy to understand that before attempting to talk about the behavior of solutions of equations such as (1.1) or (1.4) it is desirable to know that the given equation, or system of equations, has a solution. For example, the differential equation

$$y'^2 + y^2 + 1 = 0$$

clearly has no (real) solution. Both in the applications and in the theory one customarily seeks solutions of differential equations that satisfy certain additional conditions called variously *end conditions*, or *initial conditions*, or *boundary conditions*. Thus one might seek a solution $y(x)$ of the system

$$y'' + y = 0$$
$$y(0) = 0, \qquad y'(0) = 1;$$

that is, we seek a solution of the differential equation that vanishes at $x = 0$ and has the derivative 1 there. Such a solution is clearly sin x, and we shall see shortly that it is the only such solution. Again, we might seek a solution of the system

$$y'' + y = 0,$$
$$y(0) = y(\pi) = 0.$$

It may be seen at once that all functions c sin x, where c is a constant, satisfy this system. Thus, this system has an infinity of solutions. Finally, the system

$$y' = 1,$$
$$y(0) = 1, \qquad y(2) = 0$$

has no solution, for $y(x)$ must be of the form $x + c$, where c is a constant. The condition $y(0) = 1$ implies that $c = 1$. Accordingly, the solution must be $y(x) = x + 1$, but $y(2)$ cannot then be zero.

We shall employ what is known as the Picard method of successive approximations to demonstrate the existence of a solution of the following pair of differential equations:

(2.1)
$$\frac{dy}{dx} = f(x, y, z),$$

$$\frac{dz}{dx} = g(x, y, z).$$

The alterations in the statement of the theorem and in its proof that are required for the case of the general system

$$\frac{dy_1}{dx} = f_1(x, y_1, y_2, \ldots, y_n),$$

$$\frac{dy_2}{dx} = f_2(x, y_1, y_2, \ldots, y_n),$$

$$\cdot \quad \cdot \quad \cdot \quad \cdot \quad \cdot \quad \cdot \quad \cdot$$

$$\frac{dy_n}{dx} = f_n(x, y_1, y_2, \ldots, y_n) \qquad (n = 1, 2, \ldots)$$

will be evident.

To return to the differential system (2.1), we shall suppose that the functions $f(x, y, z)$ and $g(x, y, z)$ are continuous and bounded in an (open) three-dimensional region $R(x, y, z)$ and that there exist constants A and B such that

(2.2)
$$|f(x, y_1, z_1) - f(x, y_2, z_2)| \leq A|y_1 - y_2| + B|z_1 - z_2|,$$
$$|g(x, y_1, z_1) - g(x, y_2, z_2)| \leq A|y_1 - y_2| + B|z_1 - z_2|,$$

where $P_1 (x, y_1, z_1)$ and $P_2 (x, y_2, z_2)$ are any two points in R. The constants A and B are understood to be fixed constants, of course, quite independent of the points P_1 and P_2. Conditions (2.2) are called (uniform) Lipschitz conditions.

Let P_0 (x_0, y_0, z_0) be an arbitrary point of R, and let D be a rectangular domain lying in R which is defined by the inequalities

(2.3)
$$x_0 \leq x \leq x_0 + a,$$
$$y_0 - b \leq y \leq y_0 + b,$$
$$z_0 - c \leq z \leq z_0 + c,$$

where a, b, and c are positive constants. Finally, we choose a constant h that satisfies simultaneously the inequalities

$$h \leq a, \qquad h \leq \frac{b}{M}, \qquad h \leq \frac{c}{M},$$

where M is any positive constant such that both

$$|f(x, y, z)| \leq M,$$
$$|g(x, y, z)| \leq M,$$

in R.

Recall that by a solution of a system (2.1) is meant a pair (usually regarded as a column) of functions $y(x)$ and $z(x)$ of class C^1 on some interval of the x-axis which simultaneously satisfy the system on that interval.†

Theorem 2.1. Let $f(x, y, z)$ and $g(x, y, z)$ be continuous, bounded, and satisfy a uniform Lipschitz condition in an open three-dimensional region R. Then for each (x_0, y_0, z_0) in R, there exists a positive number h such that a unique solution $y(x)$, $z(x)$ of equation (2.1) exists on the interval $[x_0, x_0 + h]$ with $y(x_0) = y_0$, $z(x_0) = z_0$.

To prove the theorem, we define two sequences of functions $y_0(x)$, $y_1(x)$, ... and $z_0(x)$, $z_1(x)$, ... by means of the recursion formulas

$$y_0(x) = y_0,$$

$$y_n(x) = y_0 + \int_{x_0}^{x} f[x, y_{n-1}(x), z_{n-1}(x)] \, dx,$$

$$z_0(x) = z_0,$$

$$z_n(x) = z_0 + \int_{x_0}^{x} g[x, y_{n-1}(x), z_{n-1}(x)] \, dx \qquad (n = 1, 2, \ldots).$$

We shall show that these sequences converge, respectively, on the interval

---

† Following custom we shall use C' and C^1 interchangeably—similarly C'' and C^2, and so on.

$x_0 \leq x \leq x_0 + h$ to the functions $y(x)$ and $z(x)$ of the theorem.

First, we must show that the recursion formulas do indeed define a sequence of functions. (The principal consideration is to be sure that all our operations take place in D.) We shall show that

(2.4)
$$|y_n(x) - y_0| \leq b,$$
$$|z_n(x) - z_0| \leq c \qquad (n = 0, 1, 2, \ldots).$$

The proof is accomplished through the use of mathematical induction. Note that relations (2.4) are valid when $n = 0$. Suppose they are valid for $n = 0, 1, \ldots, r - 1$. We shall show that they are then valid for $n = r$. We have

$$|y_r(x) - y_0| = \left| \int_{x_0}^{x} f[x, y_{r-1}(x), z_{r-1}(x)] \, dx \right|$$

$$\leq \int_{x_0}^{x} |f[x, y_{r-1}(x), z_{r-1}(x)]| \, dx \leq M \int_{x_0}^{x} dx$$

$$= M(x - x_0) \leq Mh \leq b,$$

and, similarly, $|z_r(x) - z_0| \leq Mh \leq c$. Since the continuity of $y_n(x)$ and of $z_n(x)$ on the interval $x_0 \leq x \leq x_0 + h$ may also be established readily by mathematical induction, the first step in the proof is complete.

Next, consider the two infinite series

(2.5)
$$y_0(x) + [y_1(x) - y_0(x)] + [y_2(x) - y_1(x)] + \cdots,$$
$$z_0(x) + [z_1(x) - z_0(x)] + [z_2(x) - z_1(x)] + \cdots,$$

and note that the sums of the first $(n + 1)$ terms of these series are, respectively, $y_n(x)$ and $z_n(x)$. We shall show that both series converge uniformly on the interval $x_0 \leq x \leq x_0 + h$. To see this, we may verify by mathematical induction that

(2.6)
$$|y_n(x) - y_{n-1}(x)| \leq M(A + B)^{n-1} \frac{(x - x_0)^n}{n!},$$

$$|z_n(x) - z_{n-1}(x)| \leq M(A + B)^{n-1} \frac{(x - x_0)^n}{n!} \qquad (n = 1, 2, \ldots).$$

We shall indicate the proof of the former of these two inequalities. The latter may be proved in precisely similar fashion. First, we see that (as in the earlier induction proof)

$$|y_1(x) - y_0(x)| \leq M(x - x_0).$$

Suppose then that the inequalities (2.6) are valid for

$$n = 1, 2, \ldots, r - 1.$$

We shall show that they are valid when $n = r$. For,

$$|y_r(x) - y_{r-1}(x)| = \left| \int_{x_0}^x [f(x, y_{r-1}, z_{r-1}) - f(x, y_{r-2}, z_{r-2})] \, dx \right|$$

$$\leq \int_{x_0}^x [A|y_{r-1} - y_{r-2}| + B|z_{r-1} - z_{r-2}|] \, dx$$

$$\leq A \int_{x_0}^x M(A + B)^{r-2} \frac{(x - x_0)^{r-1}}{(r - 1)!} \, dx$$

$$+ B \int_{x_0}^x M(A + B)^{r-2} \frac{(x - x_0)^{r-1}}{(r - 1)!} \, dx$$

$$= M(A + B)^{r-1} \frac{(x - x_0)^r}{r!}.$$

Since the right-hand members of the inequalities (2.6) are less than

$$\frac{M(A + B)^{n-1}h^n}{n!},$$

each series in (2.5) is dominated by the series whose general term is the right-hand member in (2.6). The last series is readily seen to be a convergent positive term series. The series (2.5) then converge uniformly (and absolutely) to continuous functions $y(x)$ and $z(x)$, respectively; that is

$$\lim_{n \to \infty} y_n(x) = y(x),$$

$$\lim_{n \to \infty} z_n(x) = z(x),$$

uniformly on $x_0 \leq x \leq x_0 + h$.

The next step in the proof is to show that this pair of functions $y(x)$ and $z(x)$ provide a solution of (2.1). First, we note that from

$$y_n(x) = y_0 + \int_{x_0}^x f[x, y_{n-1}(x), z_{n-1}(x)] \, dx,$$

$$z_n(x) = z_0 + \int_{x_0}^x g[x, y_{n-1}(x), z_{n-1}(x)] \, dx,$$

we have

$$\lim_{n \to \infty} y_n(x) = y_0 + \lim_{n \to \infty} \int_{x_0}^x f[x, y_{n-1}(x), z_{n-1}(x)] \, dx$$

and

$$\lim_{n \to \infty} z_n(x) = z_0 + \lim_{n \to \infty} \int_{x_0}^{x} g[x, y_{n-1}(x), z_{n-1}(x)] \, dx.$$

But the *uniform* convergence of the sequences $\{y_n(x)\}$ and $\{z_n(x)\}$ permits us to interchange limits above, and we have

(2.7)
$$y(x) = y_0 + \int_{x_0}^{x} \lim_{n \to \infty} f[x, y_{n-1}(x), z_{n-1}(x)] \, dx,$$

$$z(x) = z_0 + \int_{x_0}^{x} \lim_{n \to \infty} g[x, y_{n-1}(x), z_{n-1}(x)] \, dx.$$

Finally, because of the continuity of f and of g we may write (2.7) in the form

(2.8)
$$y(x) = y_0 + \int_{x_0}^{x} f[x, y(x), z(x)] \, dx,$$

$$z(x) = z_0 + \int_{x_0}^{x} g[x, y(x), z(x)] \, dx.$$

The right-hand members of (2.8) possess continuous derivatives with respect to x on $x_0 \leq x \leq x_0 + h$, and, consequently, so do the left-hand members. Accordingly,

$$y'(x) = f[x, y(x), z(x)],$$
$$z'(x) = g[x, y(x), z(x)],$$

for each x on $x_0 \leq x \leq x_0 + h$. That is to say, the functions $y(x)$ and $z(x)$ satisfy (2.1) and are of class C^1 on this interval. Together, they provide a solution. We note that

(2.9)
$$y(x_0) = y_0,$$
$$z(x_0) = z_0.$$

It remains, finally, to show that the pair of functions $y(x)$ and $z(x)$ so determined will provide the *only* solution of (2.1) on the interval $x_0 \leq x \leq x_0 + h$ satisfying (2.9).

To this end let the columns

$$y(x) \qquad \bar{y}(x)$$
$$z(x) \qquad \bar{z}(x)$$

be two solutions of (2.1) on $x_0 \leq x \leq x_0 + h$ satisfying conditions (2.9). We shall show that these solutions are identical. We have

$$|y(x) - \bar{y}(x)| \leq \int_{x_0}^{x} |f[x, y(x), z(x)] - f[x, \bar{y}(x), \bar{z}(x)]|\, dx,$$

(2.10)

$$|z(x) - \bar{z}(x)| \leq \int_{x_0}^{x} |g[x, y(x), z(x)] - g[x, \bar{y}(x), \bar{z}(x)]|\, dx.$$

We shall show that the left-hand members of (2.10) are not greater than

$$2M(A + B)^{n-1} \frac{(x - x_0)^n}{n!} = t_n(x)$$

for each value of $n = 1, 2, \ldots$. Again we shall confine our attention to the difference $d = |y(x) - \bar{y}(x)|$. First, we verify the inequality for $n = 1$:

$$d \leq 2M \int_{x_0}^{x} dx = 2M(x - x_0).$$

Suppose the inequality is valid for $n = r - 1$. We apply the Lipschitz condition to the first inequality (2.10), obtaining

(2.11) $$d \leq \int_{x_0}^{x} [A|y(x) - \bar{y}(x)| + B|z(x) - \bar{z}(x)|]\, dx.$$

The induction may now be readily completed by applying the induction hypothesis to the differences in the right-hand member of (2.11).

We have then

(2.12)

$$|y(x) - \bar{y}(x)| \leq t_n(x),$$
$$|z(x) - \bar{z}(x)| \leq t_n(x)$$

for each value of x on the interval $x_0 \leq x \leq x_0 + h$ and for each value of $n = 1, 2, \ldots$.

For any fixed value of x, the left-hand members of (2.12) are definite fixed numbers greater than or equal to zero. These numbers are, however, less successively than each term in a convergent positive-term series; accordingly, they are the number zero, and

$$y(x) \equiv \bar{y}(x),$$
$$z(x) \equiv \bar{z}(x).$$

The proof of the theorem is complete.

Note that the interval $[x_0, x_0 + h]$ can be replaced by the interval $[x_0 - h', x_0]$ for a suitable choice of the positive constant h', or by $[x_0 - k, x_0 + k]$ for a suitable choice of the positive constant k.

Linear systems.   Consider the *linear system*

$$\frac{dy}{dx} = a(x)y + b(x)z = f(x, y, z),$$

(2.13)
$$\frac{dz}{dx} = c(x)y + d(x)z = g(x, y, z),$$

$$y(x_0) = y_0, \qquad z(x_0) = z_0,$$

where the functions $a(x)$, $b(x)$, $c(x)$, $d(x)$ are continuous on the interval $I : [x_0, x_0 + a]$. Let k be the maximum value of any of the functions $|a(x)|$, $|b(x)|$, $|c(x)|$, $|d(x)|$ on I. Then, inequalities (2.2) become

(2.2)′
$$|f(x, y_1, z_1) - f(x, y_2, z_2)| \le k[|y_1 - y_2| + |z_1 - z_2|],$$
$$|g(x, y_1, z_1) - g(x, y_2, z_2)| \le k[|y_1 - y_2| + |z_1 - z_2|].$$

Note that the functions defined by equations (2.13) are all continuous on I. If we take M to be any common upper bound of $|f(x, y_0, z_0)|$ and $|g(x, y_0, z_0)|$ on I and $A = B = k$, the inequalities (2.6) will hold for x on I, and we need not limit y and z as in conditions (2.3). The proof is then readily verified to hold for x on all of I.

Accordingly, if the functions $a_{ij}(x)$ are continuous on any common interval I, the unique solution of the linear system

(2.14)
$$\frac{dy_i}{dx} = a_{i1}(x)y_1 + a_{i2}(x)y_2 + \cdots + a_{in}(x)y_n(x),$$

$$y_i(x_0) = y_i^0 \qquad (i = 1, 2, \ldots, n)$$

guaranteed by the fundamental existence theorem exists on the entire interval I. In (2.14), x_0 is any point of I, and the n constants y_i^0 are arbitrary numbers. [I may be any of the types $[a, b]$, $[a, b)$, $(a, b]$, or (a, b).]

Those students familiar with complex analysis will observe that if the coefficients in the linear system are continuous in x and analytic in a parameter λ on an arbitrary λ-interval, the solution obtained by the Picard method is also analytic in λ on its interval.

Example.   Consider the differential system

$$[r(x)y']' + [p(x) + \lambda q(x)]y = 0,$$
(2.15)
$$y(a) = a_0,$$
$$y'(a) = b_0,$$

where $r(x) > 0$ and $r(x), p(x), q(x)$ are continuous on the interval I, with $a \in I$. We may set $z = r(x)y'$ and replace (2.15) by the linear system

(2.16)

$$y' = \frac{1}{r(x)}\, z,$$

$$z' = -[p(x) + \lambda q(x)]y,$$

$$y(a) = a_0, \qquad z(a) = r(a)b_0.$$

The existence theorem asserts that there exists a unique solution

$$y(x, \lambda),$$

$$z(x, \lambda)$$

of (2.16) valid on I that is of class C' in x and analytic in λ $(-\infty < \lambda < \infty)$.

Extension of solutions.   It will be observed that the proof of the existence theorem given above provides a solution that is valid on the interval $[x_0, x_0 + h]$. However, at the end point P_1 of this solution, where $x = x_0 + h$, we may apply the theorem again, inasmuch as P_1 lies in the (open) region R. We require the solution to pass through P_1 and take the interval to be $[x_0 + h - k, x_0 + h + k]$ $(k > 0)$.

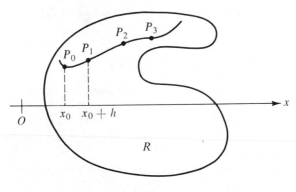

FIG. 3.1

In this fashion we obtain a sequence of "right-hand endpoints" P_1, $P_2, \ldots$ shown schematically in Fig. 3.1. This sequence either approaches the boundary of R, or has a limit point P in R. The latter alternative is not possible, for consider the solution through P. The solution must, by a similar argument, also approach the boundary of R to the left of $x = x_0$.

We shall now state the existence theorem in vector form.

Theorem 2.2. Given the vector differential system

(2.17)
$$\frac{dy}{dx} = f(x, y),$$

$$y(x_0) = y^0,$$

where $f(x, y)$ satisfies a uniform Lipschitz condition in an (open) bounded region $R(x, y)$ and the point (x_0, y^0) lies in R. There then exists a unique solution $y(x)$ of (2.17) that extends to the boundary of R.

Here $y = y_1, y_2, \ldots, y_n, y^0 = y_1^0, y_2^0, \ldots, y_n^0$, and the differential equation in (2.17) represents n first-order differential equations

$$\frac{dy_1}{dx} = f_1(x, y_1, \ldots, y_n),$$

$$\frac{dy_2}{dx} = f_2(x, y_1, \ldots, y_n),$$

$$\cdot \quad \cdot \quad \cdot \quad \cdot \quad \cdot \quad \cdot$$

$$\frac{dy_n}{dx} = f_n(x, y_1, \ldots, y_n).$$

Example.   Consider the first-order differential system

(2.18)
$$y' = 1 + y^2,$$

$$y(0) = 0.$$

Let the region R be the interior of the rectangle bounded by $x = 3, x = -3$, $y = k$, and $y = -k$, where k is an arbitrary positive number. The (unique) solution of this system is readily seen to be $\tan x$, and the curve $y = \tan x$ intersects the boundary lines $y = \pm k$ at points on the interval $(-\pi/2, \pi/2)$ however large k may be taken.

Note that $f(x, y) = 1 + y^2$ satisfies a uniform Lipschitz condition in every bounded region in the xy-plane but, unlike the situation for linear differential equations, the solution of (2.18) is valid only on the (open) interval $(-\pi/2, \pi/2)$.

We shall conclude this **section** by noting that if $f(x, y, z)$ and $g(x, y, z)$ possess in R continuous first partial derivatives with respect to both y and z, the Lipschitz condition is satisfied in every closed rectangular region $\bar{R}$ interior to R. This follows from the law of the mean:

$$f(x, y_1, z_1) - f(x, y_2, z_2) = (y_1 - y_2)f_y^* + (z_1 - z_2)f_z^*,$$

$$g(x, y_1, z_1) - g(x, y_2, z_2) = (y_1 - y_2)g_y^{**} + (z_1 - z_2)g_z^{**},$$

where the asterisks indicate evaluation of the partial derivatives at suitable mean-value points. Since there exist constants A and B such that

$$|f_y(x, y, z)| \leq A, \qquad |g_y(x, y, z)| \leq A,$$
$$|f_z(x, y, z)| \leq B, \qquad |g_z(x, y, z)| \leq B,$$

for all points (x, y, z) in $\bar{R}$, it follows at once that

$$|f(x, y_1, z_1) - f(x, y_2, z_2)| \leq A|y_1 - y_2| + B|z_1 - z_2|,$$
$$|g(x, y_1, z_1) - g(x, y_2, z_2)| \leq A|y_1 - y_2| + B|z_1 - z_2|,$$

in $\bar{R}$. This observation clearly extends to n dimensions.

3   Existence of an integrating factor

Consider the differential equation

(3.1) $$M(x, y)\, dx + N(x, y)\, dy = 0,$$

where $M(x, y)$ and $N(x, y)$ are functions of class C' in an open square region R and $N(x, y) \neq 0$ there. We may then write equation (3.1) in the form

(3.2) $$y' = -\frac{M(x, y)}{N(x, y)}.$$

If $P_0(x_0, y_0)$ is any point of R, by the fundamental existence theorem there exists a unique solution

(3.3) $$y = A(x, y_0)$$

of (3.2) near P_0 with the property that

(3.4) $$A(x_0, y_0) \equiv y_0,$$

identically in y_0.

We shall show next that near P_0 equation (3.3) can be solved for y_0. To do this we shall appeal to the implicit-function theorem. First, it follows from equation (3.4) that P_0 is a solution point of equation (3.3). Next, we may evaluate

$$\frac{\partial A}{\partial y_0}\bigg|^{P_0}$$

by differentiating (3.4) with respect to y_0. We then have

$$\frac{\partial A(x_0, y_0)}{\partial y_0} = 1.$$

Because P_0 is a solution point and $\partial A/\partial y_0 \neq 0$ at P_0, equation (3.3) has a solution

(3.5) $$y_0 = \varphi(x, y)$$

which is valid neighboring P_0. It follows that

(3.6) $$y \equiv A[x, \varphi(x, y)],$$

and that along an integral curve (3.5)

$$\frac{dy}{dx} = -\frac{\varphi_x(x, y)}{\varphi_y(x, y)},$$

provided $\varphi_y \neq 0$. To show that $\varphi_y \neq 0$ we may differentiate (3.6) partially with respect to y obtaining

$$1 \equiv \frac{\partial A[x, \varphi(x, y)]}{\partial y_0} \frac{\partial \varphi}{\partial y}.$$

Inasmuch as

$$\frac{\partial A[x_0, \varphi(x_0, y_0)]}{\partial y_0} = \frac{\partial A(x_0, y_0)}{\partial y_0} = 1,$$

it follows that $\varphi_y = 1$ at P_0 and, therefore, that $\varphi_y \neq 0$ near P_0.

Accordingly, along an integral curve of the family $\varphi(x, y) = y_0$ both

$$\frac{dy}{dx} = -\frac{\varphi_x(x, y)}{\varphi_y(x, y)}$$

and

$$\frac{dy}{dx} = -\frac{M(x, y)}{N(x, y)}.$$

Thus,

$$\varphi_x(x, y) = \frac{\varphi_y(x, y)}{N(x, y)} M(x, y),$$

$$\varphi_y(x, y) = \frac{\varphi_y(x, y)}{N(x, y)} N(x, y).$$

We have proved, then, the existence of an integrating factor

(3.7) $$I(x, y) = \frac{\varphi_y(x, y)}{N(x, y)}$$

for the differential equation (3.1). Note that $I(x, y) \neq 0$ near P_0. Accordingly, near P_0 the multiplication of both members of (3.1) by $I(x, y)$ does not introduce extraneous solutions.

As an application, let us use this result to derive an integrating factor for the homogeneous differential equation

(3.8) $$M(x, y)\, dx + N(x, y)\, dy = 0,$$

where

(3.9) $$\frac{M(ax, ay)}{N(ax, ay)} \equiv \frac{M(x, y)}{N(x, y)}.$$

For definiteness, we suppose that $M(x, y)$ and $N(x, y)$ are of class C' in a region R, and that $N(x, y) \neq 0$ in R. Further, we suppose for simplicity that R does not contain the origin, but does contain the point $(1, 1)$.

The substitution

$$y = vx, \qquad dy = v\, dx + x\, dv,$$

in (3.8) leads, as we have seen [Chapter 1, Section 5], to the differential equation

$$[M(1, v) + vN(1, v)]\, dx + xN(1, v)\, dv = 0.$$

Accordingly, the equation

(3.10) $$\varphi(x, y) \equiv \int_1^x \frac{dx}{x} + \int_1^{y/x} \frac{N(1, v)\, dv}{M(1, v) + vN(1, v)} = c$$

provides integral curves of equation (3.8). We readily compute $\varphi_y(x, y)$ from (3.10), and we have

$$y = \frac{1}{x} \left[\frac{N\left(1, \frac{y}{x}\right)}{M\left(1, \frac{y}{x}\right) + \frac{y}{x} N\left(1, \frac{y}{x}\right)} \right]$$

$$= \frac{N\left(1, \frac{y}{x}\right)}{xM\left(1, \frac{y}{x}\right) + yN\left(1, \frac{y}{x}\right)}$$

$$= \frac{N(x, y)}{xM(x, y) + yN(x, y)}.$$

The last equality follows from (3.9) when a is set equal to $1/x$.

According to (3.7),

$$I(x, y) = \frac{\varphi_y(x, y)}{N(x, y)} = \frac{1}{xM(x, y) + yN(x, y)}$$

—a result which is valid unless

$$\frac{M(x, y)}{N(x, y)} \equiv -\frac{y}{x};$$

that is, unless the given differential equation can be written as

$$y \, dx - x \, dy = 0.$$

Exercises

Solve the following differential equations and then employ equation (3.7) to determine an integrating factor for each equation.

1. $x \, dy - y \, dx = 0.$

2. $(x - y) \, dx + dy = 0.$

3. $(x - y) \, dx + (x + 2y) \, dy = 0.$

4. $[a(x)y - f(x)] \, dx + dy = 0$, where $a(x)$ and $f(x)$ are continuous on an interval $a < x < b$.

5. $3xy \, dx + (x^2 + y^2) \, dy = 0.$

4

Linear differential equations: elementary theory

In this chapter we shall be concerned with the elementary theory of second-order linear differential equations. The second-order theory has considerable intrinsic significance. Further, the extension of this elementary theory to linear differential equations of order n $(n > 2)$ is usually immediate and quite clear. Accordingly, the ideas developed below have, in this sense, double significance.

1 Definitions; the fundamental existence theorem

By a linear differential equation of order n is meant an equation of the form

(1.1) $\qquad a_0(x)y^{(n)} + a_1(x)y^{(n-1)} + \cdots + a_{n-1}(x)y' + a_n(x)y = f(x),$

where $a_0(x) \not\equiv 0$. In order that we may have a reasonable amount of continuity at our disposal we shall always suppose that the functions $a_0(x)$, $a_1(x), \ldots, a_n(x)$, $f(x)$ are continuous on an interval I and that $a_0(x) \neq 0$ on this interval. The interval I may be either of finite or of infinite length and may be open or closed at either end. It will be recalled that a solution $y(x)$ over I will have the property, according to our earlier definition, that it together with its derivatives $y'(x)$, $y''(x)$, $\ldots$, $y^{(n-1)}(x)$ will be continuous on this interval. It follows from (1.1) that $y^{(n)}(x)$ will then also be continuous on the interval.

We state next the fundamental existence theorem as it applies to (1.1).

Theorem 1.1. Let $x = x_0$ be a point of the interval I and let $c_0, c_1, c_2, \ldots, c_{n-1}$ be an arbitrary set of n real numbers. There then exists one and only one solution $y(x)$ of (1.1) with the property that

$$y(x_0) = c_0, \qquad y'(x_0) = c_1, \qquad \ldots, \qquad y^{(n-1)}(x_0) = c_{n-1}.$$

Further, this solution is defined over the entire interval I.

Thus, at any point $x = x_0$ of the interval we may prescribe the values $y(x_0), y'(x_0), \ldots, y^{(n-1)}(x_0)$ quite arbitrarily, and there will then exist precisely one solution over the whole interval which at the point $x = x_0$ assumes the prescribed values.

For example, suppose the interval I is the interval $0 \le x \le 1$ and that $n = 3$. Then at the point $x = \frac{1}{3}$ we may prescribe that $y = \sqrt{2}$, $y' = \pi^2$, $y'' = 10^{10}$, and the theorem asserts the existence of one and exactly one solution $y(x)$ taking on these values at $x = \frac{1}{3}$. This solution will, further, be defined at every point of the interval $0 \le x \le 1$, and $y(x)$, $y'(x)$, $y''(x)$, and $y'''(x)$ will be continuous functions at each point of this interval.

Recall that a function $f(x)$ which together with its first n derivatives is continuous on an interval is said to be of class C^n on this interval. Thus, a solution of (1.1) is by definition of class C^{n-1}, and it follows from (1.1) that it is also of class C^n on I.

It is important to note that if we discover (no matter how) a solution $y(x)$ on I that satisfies the conditions

$$y(x_0) = c_0, \qquad y'(x_0) = c_1, \qquad \ldots, \qquad y^{(n-1)}(x_0) = c_{n-1},$$

we have found *the* solution we seek. For $y(x)$ will obviously be *a* solution of our problem, but by Theorem 1.1 there is exactly one such solution; hence, $y(x)$ is *the* solution.

If $f(x) \equiv 0$ on I, equation (1.1) becomes

(1.2) $\qquad a_0(x)y^{(n)} + a_1(x)y^{(n-1)} + \cdots + a_{n-1}(x)y' + a_n(x)y = 0.$

We say then that (1.2) is *homogeneous*.

Corollary. A solution $y(x)$ of (1.2) with the property that at a point $x = x_0$ of I

(1.3) $\qquad y(x_0) = 0, \qquad y'(x_0) = 0, \qquad \ldots, \qquad y^{(n-1)}(x_0) \doteq 0$

is identically zero on I.

The proof of the corollary is extremely simple. We note that $y(x) = 0$ is first of all a solution of (1.2) which satisfies (1.3). By the uniqueness guaranteed by the fundamental existence theorem it is the *only* solution with this property. The proof is complete.

The student will readily verify that if $y_1(x)$ and $y_2(x)$ are any two solutions of (1.2), the *linear combination*† $c_1 y_1(x) + c_2 y_2(x)$, where c_1 and c_2 are constants, is also a solution of (1.2). In the same fashion it can readily be seen that a linear combination of any number of solutions of (1.2) is a solution of (1.2).

Exercises

Prove that if $y_1(x)$ and $y_2(x)$ are solutions of (1.2), so is $c_1 y_1(x) + c_2 y_2(x)$, where c_1 and c_2 are arbitrary constants.

2   Linear dependence and linear independence

The concepts of *linear dependence* and *linear independence* are basic in what follows. We met these ideas in Chapter 2, but we shall repeat their definitions here. Two solutions $u_1(x)$ and $u_2(x)$ are said to be *linearly dependent* if there exist constants c_1 and c_2, not both zero, such that

$$c_1 u_1(x) + c_2 u_2(x) \equiv 0 \qquad (x \text{ on } I).$$

If $u_1(x)$ and $u_2(x)$ are not linearly dependent, they are said to be *linearly independent*.

The student should note that if either solution is identically zero, the two solutions are linearly dependent. For, suppose $u_1(x) \equiv 0$. Then taking $c_1 = 1, c_2 = 0$ we have $c_1 u_1(x) + c_2 u_2(x) \equiv 0$.

Theorem 2.1. There exist two linearly independent solutions $y_1(x)$ and $y_2(x)$ of the equation

(2.1) $$a_0(x)y'' + a_1(x)y' + a_2(x)y = 0$$

with the property that every solution $y(x)$ of (2.1) may be written in the form

(2.2) $$y(x) = c_1 y_1(x) + c_2 y_2(x),$$

where c_1 and c_2 are suitably chosen constants.

---

† Recall the footnote on p. 4.

Before we prove Theorem 2.1, let us recall that we have assumed that $a_0(x)$, $a_1(x)$, and $a_2(x)$ are continuous and $a_0(x) \neq 0$ on the interval I.

To prove the theorem, let $x = x_0$ be a point of I and let $y_1(x)$ and $y_2(x)$ be the solutions (guaranteed by the fundamental existence theorem) with the properties

(2.3)
$$y_1(x_0) = 1, \qquad y_2(x_0) = 0,$$
$$y_1'(x_0) = 0, \qquad y_2'(x_0) = 1.$$

We shall first prove that $y_1(x)$ and $y_2(x)$ are linearly independent. Let c_1 and c_2 be any constants such that

(2.4)
$$c_1 y_1(x) + c_2 y_2(x) \equiv 0,$$

and, hence, such that

(2.5)
$$c_1 y_1'(x) + c_2 y_2'(x) \equiv 0$$

on I. The identities (2.4) and (2.5) will hold, in particular, when $x = x_0$; that is,

(2.6)
$$c_1 y_1(x_0) + c_2 y_2(x_0) = 0,$$
$$c_1 y_1'(x_0) + c_2 y_2'(x_0) = 0.$$

Applying (2.3) to (2.6), we see that $c_1 = 0$ and $c_2 = 0$. Accordingly, $y_1(x)$ and $y_2(x)$ are linearly independent.

To complete the proof of the theorem, let $y(x)$ be an arbitrary solution of (2.1). We wish to show that $y(x)$ may be written as a linear combination of $y_1(x)$ and $y_2(x)$. To that end consider the solution

$$Y(x) = y(x) - y(x_0)y_1(x) - y'(x_0)y_2(x).$$

Note that $Y(x)$ *is*, indeed, a solution, for it is a linear combination of solutions. We shall show that $Y(x)$ is a solution such that $Y(x_0) = 0$, $Y'(x_0) = 0$. It will follow from the corollary to Theorem 1.1 that $Y(x) \equiv 0$, and the proof will be complete. Note that, by equations (2.3),

(2.7)
$$Y(x_0) = y(x_0) - y(x_0)y_1(x_0) - y'(x_0)y_2(x_0) = 0,$$
$$Y'(x_0) = y'(x_0) - y(x_0)y_1'(x_0) - y'(x_0)y_2'(x_0) = 0.$$

The proof is complete. The "suitably chosen constants" referred to in the statement of the theorem have the values

$$c_1 = y(x_0), \qquad c_2 = y'(x_0).$$

Theorem 2.1 permits us quite properly to call $c_1 y_1(x) + c_2 y_2(x)$, where c_1 and c_2 are constants, the *general solution* of (2.1). For every such linear combination of solutions is a solution of (2.1) and, conversely, every solution of (2.1) may be put in this form.

If $y_1(x)$ and $y_2(x)$ are any two solutions of (2.1), the determinant

$$w(x) = \begin{vmatrix} y_1(x) & y_2(x) \\ y_1'(x) & y_2'(x) \end{vmatrix} = y_1(x)y_2'(x) - y_2(x)y_1'(x),$$

known as the *wronskian*† of the two solutions, enables us to decide very simply whether or not the solutions $y_1(x)$ and $y_2(x)$ are linearly dependent.

Theorem 2.2. Two solutions $y_1(x)$ and $y_2(x)$ of (2.1) are linearly dependent if and only if their wronskian is identically zero.

We shall prove the sufficiency of the condition first; that is, we shall suppose $w(x)$ to be identically zero and prove that the solutions are linearly dependent. We have then

$$(2.8) \qquad \begin{vmatrix} y_1(x) & y_2(x) \\ y_1'(x) & y_2'(x) \end{vmatrix} \equiv 0.$$

It follows that if $x = x_0$ is any point of I,

$$(2.9) \qquad \begin{vmatrix} y_1(x_0) & y_2(x_0) \\ y_1'(x_0) & y_2'(x_0) \end{vmatrix} = 0.$$

There then exist constants c_1 and c_2, not both zero, such that‡

$$(2.10) \qquad \begin{aligned} c_1 y_1(x_0) + c_2 y_2(x_0) &= 0, \\ c_1 y_1'(x_0) + c_2 y_2'(x_0) &= 0. \end{aligned}$$

Consider the solution $c_1 y_1(x) + c_2 y_2(x)$. According to (2.10) this solution and its derivative are zero when $x = x_0$; hence, $c_1 y_1(x) + c_2 y_2(x) \equiv 0$. The proof of the sufficiency is complete.

To prove the necessity of the condition, we assume that there exist constants c_1 and c_2, not both zero, such that

$$(2.11) \qquad c_1 y_1(x) + c_2 y_2(x) \equiv 0.$$

We wish to prove that the wronskian $w(x) \equiv 0$. From (2.11) we have

$$(2.12) \qquad c_1 y_1'(x) + c_2 y_2'(x) \equiv 0.$$

Let $x = x_0$ be an arbitrary point of I. The identities (2.11) and (2.12)

---

† $-w(x)$ is also known as "the" wronskian of $y_1(x)$ and $y_2(x)$. The apparent ambiguity will not present any difficulty in what follows.

‡ The student will recall that a system of n homogeneous linear algebraic equations in n unknowns has solutions, other than the null solution, if and only if the determinant of the coefficients vanishes.

hold, in particular, when $x = x_0$. There thus exist constants c_1 and c_2, not both zero, such that equations (2.10) hold. It follows that the determinant of the coefficients of c_1 and c_2 must be zero; that is, (2.9) must hold. But $x = x_0$ was an arbitrary point of I, and (2.8) must hold throughout I.

The proof of the theorem is complete.

Theorem 2.3. The wronskian of two solutions is either identically zero or never zero on I.

First, let us remind ourselves that the interval I may be either open or closed at either end point of the interval, but in all cases it is an interval in which the functions $a_0(x)$, $a_1(x)$, and $a_2(x)$ of equation (2.1) are continuous with $a_0(x) \neq 0$.

To prove the theorem, note that

$$w'(x) = \begin{vmatrix} y_1'(x) & y_2'(x) \\ y_1'(x) & y_2'(x) \end{vmatrix} + \begin{vmatrix} y_1(x) & y_2(x) \\ y_1''(x) & y_2''(x) \end{vmatrix}$$

$$= y_1(x)y_2''(x) - y_2(x)y_1''(x)$$

$$= y_1(x)\left[-\frac{a_1}{a_0}y_2'(x) - \frac{a_2}{a_0}y_2(x) \right] - y_2(x)\left[-\frac{a_1}{a_0}y_1'(x) - \frac{a_2}{a_0}y_1(x) \right]$$

$$= -\frac{a_1}{a_0}[y_1(x)y_2'(x) - y_2(x)y_1'(x)]$$

$$= -\frac{a_1}{a_0}w(x).$$

Thus, $w(x)$ is a solution of the first-order homogeneous linear differential equation

$$w' + \frac{a_1}{a_0}w = 0,$$

and according to the corollary in Section 1, $w(x) \equiv 0$ if it is zero at any point of I.

We conclude this section with the following result.

Theorem 2.4. If $y_1(x)$ and $y_2(x)$ are any pair of linearly independent solutions of equation (2.1), every solution $y(x)$ of (2.1) may be written in the form

$$y(x) = c_1 y_1(x) + c_2 y_2(x),$$

where c_1 and c_2 are suitably chosen constants.

The proof is left to the student as an exercise.

It will be observed that every theorem in this section, as well as its proof, is readily extended to linear differential equations of order n.

Exercises

1. Show that $\sin x$ and $\cos x$ are linearly independent solutions of $y'' + y = 0$.

2. Show that $y_1(x) = \sin x$ and $y_2(x) = \sin x - \cos x$ are linearly independent solutions of $y'' + y = 0$. Determine constants c_1 and c_2 so that the solution

$$\sin x + 3 \cos x \equiv c_1 y_1(x) + c_2 y_2(x).$$

3. Find two linearly independent solutions of the equation

$$x^2 y'' + xy' + 4y = 0 \qquad (x > 0).$$

4. Find the solution of the differential system

$$x^2 y'' - xy' + 2y = 0,$$
$$y(-1) = 0,$$
$$y'(-1) = 1.$$

5. Show that linearly independent solutions of

$$y'' - 2y' + 2y = 0$$

are $e^x \sin x$ and $e^x \cos x$. What is the general solution? Find the solution $y(x)$ with the property that $y(0) = 2$, $y'(0) = -3$. Find another pair of linearly independent solutions.

6. Show that if a is a positive constant, linearly independent solutions of $y'' + a^2 y = 0$ are $\sin ax$ and $\cos ax$. What is the general solution?

7. Find the general solution of the system

$$y'' + 4y = 0,$$
$$y'(0) = 0.$$

Show that every solution $y(x)$ of this system vanishes when $x = \pi/4$.

8. Show that on the interval $0 < x < \infty$, $\sin \dfrac{1}{x}$ and $\cos \dfrac{1}{x}$ are linearly independent solutions of the equation $x^4 y'' + 2x^3 y' + y = 0$. Note that $x = 0$ is a singular point of the differential equation. Sketch the former solution on the interval $0 < x < \dfrac{1}{\pi}$. Find the solution $y(x)$ of the differential equation with the property that $y\left(\dfrac{1}{\pi}\right) = 1$, $y'\left(\dfrac{1}{\pi}\right) = -1$.

9.   Show that $\sin x^2$ and $\cos x^2$ are linearly independent solutions of the differential equation $xy'' - y' + 4x^3y = 0$. Show that their wronskian has a zero at $x = 0$, but is not identically zero. Why does this not contradict Theorem 2.3?

10. Prove Theorem 2.4.

11. Given the differential equation

$$a_0(x)y'' + a_1(x)y' + a_2(x)y = 0$$

with the usual conditions on $a_i(x)$, show that a necessary and sufficient condition that two solutions $y_1(x)$ and $y_2(x)$ be linearly dependent is that one solution be a constant times the other.

12. Refer to Exercise 11 and show that if $y_0(x) \not\equiv 0$ is any solution of the differential equation with the property that $y_0(c) = 0$ $(a \le c \le b)$, then every solution $y(x)$ linearly dependent on $y_0(x)$ vanishes when $x = c$. Conversely, if $y_0(x)$ and $y(x)$ vanish at $x = c$, then $y_0(x)$ and $y(x)$ are linearly dependent.

13. Prove that if any nonnull solution of the system

$$a_0(x)y'' + a_1(x)y' + a_2(x)y = 0,$$
$$y(x_0) = 0$$

vanishes when $x = c$ $(x_0 < c \le b)$, all solutions of this system vanish when $x = c$. Show similarly that if a solution has no zero except $x = x_0$, all solutions of the system have this property. Assume the usual conditions on the coefficients $a_i(x)$.

14. Prove that if any nonnull solution of the system

$$a_0(x)y'' + a_1(x)y' + a_2(x)y = 0,$$
$$y'(x_0) = 0$$

vanishes when $x = c$ $(x_0 < c \le b)$, all solutions of this system vanish when $x = c$. Show similarly that if one solution of the system never vanishes, all solutions of the system have this property. Assume the usual conditions on the coefficients $a_i(x)$.

15. Define solutions $s(x)$ and $c(x)$ of the differential equation $y'' + y = 0$ by the conditions $s(0) = 0$, $s'(0) = 1$, $c(0) = 1$, $c'(0) = 0$. Prove the well-known theorems (a) $s'(x) \equiv c(x)$; (b) $c'(x) \equiv -s(x)$; (c) $s^2(x) + c^2(x) \equiv 1$, without use of trigonometry.

16. Refer to Exercise 15 and without use of trigonometry prove that

$$s(x + a) = s(x)c(a) + c(x)s(a),$$
$$c(x + a) = c(x)c(a) - s(x)s(a).$$

17.   Refer to Exercise 16 and without use of trigonometry prove that

$$s(2x) = 2s(x)c(x),$$
$$c(2x) = c^2(x) - s^2(x).$$

Answers

3. $\sin \log x^2$, $\cos \log x^2$.

4. $x \sin \log (-x)$.

3 Variation of parameters

We now turn our attention to the differential equation

(3.1) $a_0(x)y'' + a_1(x)y' + a_2(x)y = f(x),$

where, as usual, $a_0(x)$, $a_1(x)$, $a_2(x)$, and $f(x)$ are continuous with $a_0(x) \neq 0$ on an interval I. With (3.1) we associate the *corresponding* homogeneous equation

(3.2) $a_0(x)y'' + a_1(x)y' + a_2(x)y = 0.$

Equation (3.2), as we have seen, has a general solution in the form

(3.3) $c_1 y_1(x) + c_2 y_2(x),$

where $y_1(x)$ and $y_2(x)$ are linearly independent solutions of (3.2) and c_1 and c_2 are arbitrary constants. We are now prepared to prove the following result concerning the general solution of (3.1).

Theorem 3.1. If $y_0(x)$ is any one solution of (3.1), the general solution of (3.1) may be written in the form

$$y_0(x) + c_1 y_1(x) + c_2 y_2(x),$$

where $y_1(x)$ and $y_2(x)$ are linearly independent solutions of (3.2) and c_1 and c_2 are arbitrary constants.

Thus, to find the general solution of (3.1), we first find the general solution of (3.2). Next, we determine a particular solution of (3.1) by some convenient method. The sum of the two is the general solution of (3.1).

Before proving the theorem let us illustrate it. Consider the differential equation

(3.1)' $y'' + y = 1.$

The associated homogeneous equation is

$$(3.2)' \qquad\qquad y'' + y = 0.$$

The general solution of $(3.2)'$ is

$$c_1 \sin x + c_2 \cos x.$$

By inspection, we observe that $y = 1$ is a particular solution of $(3.1)'$. Thus, the general solution of $(3.1)'$ is

$$1 + c_1 \sin x + c_2 \cos x,$$

where c_1 and c_2 are arbitrary constants.

We return now to the proof of the theorem. Suppose that $y(x)$ is an arbitrary solution of (3.1), and consider the function $Y(x) = y(x) - y_0(x)$. By hypothesis, we have

$$a_0 y''(x) + a_1 y'(x) + a_2 y(x) \equiv f(x),$$
$$a_0 y_0''(x) + a_1 y_0'(x) + a_2 y_0(x) \equiv f(x).$$

Subtracting these two identities, we have

$$a_0(y - y_0)'' + a_1(y - y_0)' + a_2(y - y_0) \equiv 0.$$

Thus, $Y(x) = y(x) - y_0(x)$ is a solution of (3.2) and may therefore be written in the form

$$Y(x) \equiv c_1 y_1(x) + c_2 y_2(x).$$

That is,

$$y(x) \equiv y_0(x) + c_1 y_1(x) + c_2 y_2(x),$$

as was to have been proved.

The extension of Theorem 3.1 to linear differential equations of order n is immediate.

Ordinarily we cannot hope to find, by inspection, the one solution of (3.1) that we need. Fortunately, if we can solve (3.2), the method of *variation of parameters* always produces a particular solution of (3.1). This result may be established as follows. Suppose $y_1(x)$ and $y_2(x)$ are linearly independent solutions of the corresponding homogeneous equation (3.2). We first try to determine functions $u_1(x)$ and $u_2(x)$ such that

$$y_0(x) = u_1(x) y_1(x) + u_2(x) y_2(x)$$

is a particular solution of (3.1) and such that

$$(3.4) \qquad\qquad u_1'(x) y_1(x) + u_2'(x) y_2(x) = 0.$$

[It is the auxiliary condition (3.4) which leads to the name "variation of parameters."] To that end substitute $y_0(x)$ in (3.1), and, using (3.4), obtain

(3.5) $$u_1'(x)y_1'(x) + u_2'(x)y_2'(x) = \frac{f(x)}{a_0(x)}.$$

Equations (3.4) and (3.5) may be solved for $u_1'(x)$ and $u_2'(x)$:

(3.6) $$u_1'(x) = -\frac{y_2(x)f(x)}{a_0(x)w(x)}, \qquad u_2'(x) = \frac{y_1(x)f(x)}{a_0(x)w(x)},$$

where $w(x) = y_1(x)y_2'(x) - y_1'(x)y_2(x)$ is the wronskian of $y_1(x)$ and $y_2(x)$ and is never zero. From equations (3.6) we have

$$u_1(x) = -\int_{x_0}^x \frac{y_2(t)f(t)}{a_0(t)w(t)}\,dt, \qquad u_2(x) = \int_{x_0}^x \frac{y_1(t)f(t)}{a_0(t)w(t)}\,dt.$$

The point $x = x_0$ may be any convenient point of I. A particular solution of (3.1) is then

$$y_0(x) = u_1(x)y_1(x) + u_2(x)y_2(x),$$

which may be written in the symmetric form

(3.7) $$y_0(x) = \int_{x_0}^x \begin{vmatrix} y_1(t) & y_2(t) \\ y_1(x) & y_2(x) \end{vmatrix} \frac{f(t)}{a_0(t)w(t)}\,dt.$$

It will be noted that $w(t) \neq 0$ on I.

Example. Using variation of parameters, find the general solution of

(3.8) $$y'' + y = x.$$

Here the general solution of the corresponding homogeneous equation is $c_1 \cos x + c_2 \sin x$, and $f(x) = x$, $a_0(x) = 1$, $w(x) = 1$. Equations (3.4) and (3.5) become

$$u_1'(x) \cos x + u_2'(x) \sin x = 0,$$
$$u_1'(x)(-\sin x) + u_2'(x) \cos x = x.$$

Hence,

$$u_1'(x) = -x \sin x, \qquad u_2'(x) = x \cos x,$$

and $u_1(x)$ and $u_2(x)$ may be taken as

$$u_1(x) = x \cos x - \sin x, \qquad u_2(x) = x \sin x + \cos x.$$

A particular solution $y_0(x) = u_1 y_1 + u_2 y_2$ is then $y_0(x) = x$. The general solution is $x + c_1 \cos x + c_2 \sin x$.

In this example we take $x_0 = 0$, $y_1(x) = \cos x$, $y_2(x) = \sin x$, $w(x) = 1$, and (3.7) becomes

$$y_0(x) = \int_0^x \begin{vmatrix} \cos t & \sin t \\ \cos x & \sin x \end{vmatrix} t\, dt$$

$$= \int_0^x t \sin (x - t)\, dt$$

$$= x - \sin x.$$

This choice for $y_0(x)$ would have yielded the same answers as before. In this instance, we could also have easily guessed the particular solution x by inspection of (3.8).

Example.   Guessing a particular solution of the differential equation

(3.8)′ $$y'' + y = \csc x \qquad (0 < x < \pi)$$

would be more difficult. In this example, we take $x_0 = \pi/2$, $y_1(x) = \cos x$, $y_2(x) = \sin x$, $w(x) = 1$, and (3.7) becomes

$$y_0(x) = \int_{\pi/2}^x \begin{vmatrix} \cos t & \sin t \\ \cos x & \sin x \end{vmatrix} \csc t\, dt$$

$$= \int_{\pi/2}^x \csc t(\sin x \cos t - \cos x \sin t)\, dt$$

$$= \sin x(\log \sin x) - x \cos x + \frac{\pi}{2} \cos x.$$

The general solution of (3.8)′ may then be written as

$$\sin x(\log \sin x) - x \cos x + c_1 \sin x + c_2 \cos x.$$

The general case.   Consider the differential equation

(3.9) $$a_0(x)y^{(n)} + a_1(x)y^{(n-1)} + \cdots + a_{n-1}(x)y' + a_n(x)y = f(x),$$

where the functions $a_i(x)$ and $f(x)$ are continuous on an interval I, with $a_0(x) \neq 0$ there. Let linearly independent solutions of the homogeneous equation corresponding to (3.9) be $y_1(x), y_2(x), \ldots, y_n(x)$. As in the second-order case, we seek a particular solution $y_0(x)$ of (3.9) of the form

(3.10) $$y_0(x) = u_i(x)y_i(x) \qquad (i = 1, 2, \ldots, n).$$

In the right-hand member of (3.10), for convenience, we are using the convention of tensor analysis that a repeated subscript (or superscript) indicates summation with respect to that subscript. Thus,

$$u_i(x)y_i(x) = \sum_{i=1}^{n} u_i(x)y_i(x).$$

Next, we impose on the functions $u_i(x)$ the $(n-1)$ conditions

(3.11)
$$u_i'y_i = 0,$$
$$u_i'y_i' = 0,$$
$$\cdot \quad \cdot \quad \cdot \quad \cdot$$
$$u_i'y_i^{(n-2)} = 0.$$

It follows that

(3.12)
$$y_0 = u_iy_i,$$
$$y_0' = u_iy_i',$$
$$y_0'' = u_iy_i'',$$
$$\cdot \quad \cdot \quad \cdot \quad \cdot \quad \cdot \quad \cdot \quad \cdot$$
$$y_0^{(n)} = u_iy_i^{(n)} + u_i'y_i^{(n-1)}$$

We substitute (3.12) in (3.9), obtaining

$$u_i[a_0y_i^{(n)} + a_1y_i^{(n-1)} + \cdots + a_ny_i] + a_0u_i'y_i^{(n-1)} = f,$$

or

(3.13)
$$a_0u_i'y_i^{(n-1)} = f,$$

inasmuch as the quantity in the bracket is zero. Equations (3.11) combined with (3.13) then yield the n equations

(3.14)
$$u_1'y_1 + u_2'y_2 + \cdots + u_n'y_n = 0,$$
$$u_1'y_1' + u_2'y_2' + \cdots + u_n'y_n' = 0,$$
$$\cdot \quad \cdot \quad \cdot \quad \cdot \quad \cdot \quad \cdot \quad \cdot \quad \cdot \quad \cdot \quad \cdot$$
$$u_1'y_1^{(n-2)} + u_2'y_2^{(n-2)} + \cdots + u_n'y_n^{(n-2)} = 0,$$
$$u_1'y_1^{(n-1)} + u_2'y_2^{(n-1)} + \cdots + u_n'y_n^{(n-1)} = \frac{f}{a_0}.$$

Equations (3.14) can be solved uniquely for $u_1', u_2', \ldots, u_n'$, inasmuch as the determinant of the coefficients is the wronskian $w(x)$ of the linearly independent solutions $y_1, y_2, \ldots, y_n$ of the homogeneous linear differential equation. An easy computation now yields

$$
(3.15) \quad y_0(x) = \int_{x_0}^{x}
\begin{vmatrix}
y_1(t) & y_2(t) & \cdots & y_n(t) \\
y_1'(t) & y_2'(t) & \cdots & y_n'(t) \\
\cdot & \cdot & \cdots & \cdot \\
y_1^{(n-2)}(t) & y_2^{(n-2)}(t) & \cdots & y_n^{(n-2)}(t) \\
y_1(x) & y_2(x) & \cdots & y_n(x)
\end{vmatrix}
\frac{f(t)}{a_0(t)w(t)}\,dt,
$$

where x_0 is an arbitrary point of I.

Exercises

1.   Find the general solution of $y'' + y = 2 - x$ both by guessing a particular solution and by finding a particular solution by variation of parameters.

2.   Find a solution as in Exercise 1 for $y'' + 4y = e^x$.

3.   Using the method of variation of parameters, find the general solution of $y'' + y = \tan x$.

4.   Using the method of variation of parameters, find the general solution of $y'' + y = \sec x$.

5.   Use the method of variation of parameters to find the general solution of $y'' + y = 2 \sin x$.

6.   Prove that (3.7) provides a solution of (3.1) by substituting the solution in (3.1). The reversibility of the steps leading to (3.7) will be seen to provide an alternate proof.

Answers

1. $2 - x + c_1 \sin x + c_2 \cos x$.

3. $c_1 \sin x + c_2 \cos x - \cos x \log |\sec x + \tan x|$.

4. $c_1 \sin x + c_2 \cos x + x \sin x + \cos x \log |\cos x|$.

5. $c_1 \sin x + c_2 \cos x - x \cos x$.

4   The adjoint equation

In Chapter 1 we verified that

$$
(4.1) \qquad z(x) = \frac{1}{a_0(x)}\, e^{\int [a_1(x)/a_0(x)]\,dx}
$$

is an integrating factor of the first-order linear differential equation

(4.2) $$a_0(x)y' + a_1(x)y = f(x).$$

This integrating factor might have been obtained by seeking a function $z(x)$ such that

(4.3) $$z(x)[a_0(x)y' + a_1(x)y] \equiv \frac{d}{dx}[r(x)y],$$

where we assume that $a_0(x)$, $a_0'(x)$, $a_1(x)$ are continuous and $a_0(x) \neq 0$ on an interval I. The function $r(x)$ is also to be determined. Clearly, from (4.3), we desire to determine $z(x)$ and $r(x)$ so that

(4.4) $$z(x)a_0(x) = r(x), \qquad r'(x) = z(x)a_1(x).$$

These equations lead at once to the differential equation

(4.5) $$[z(x)a_0(x)]' - z(x)a_1(x) = 0$$

for the determination of $z(x)$. Equation (4.5) is a first-order linear homogeneous equation, a solution of which is given by (4.1). The function $r(x)$ is then given by the first of equations (4.4).

If we write

$$L_1(y) = a_0(x)y' + a_1(x)y,$$
$$M_1(z) = -(a_0 z)' + (a_1 z),$$

$M_1(z)$ is said to be the *adjoint* of $L_1(y)$. The differential equation $L_1(y) = 0$, that is,

$$a_0(x)y' + a_1(x)y = 0,$$

is said to be *self-adjoint* if $L_1(y) \equiv -M_1(y)$. This will be true if

$$a_0'(x) = 2a_1(x);$$

that is, if the differential equation $L_1(y) = 0$ can be written in the form

$$a_0(x)y' + \tfrac{1}{2}a_0'(x)y = 0.$$

The equation $L_1(y) = 0$ can always be put in self-adjoint form by writing it as

$$q(x)y' + \frac{a_1(x)q(x)}{a_0(x)}y = 0,$$

where

$$q(x) = e^{2\int \frac{a_1(x)}{a_0(x)}dx}.$$

In general, every second-order linear differential equation can also be put in self-adjoint form. [This is not true for general linear differential equations of order higher than the second.] We can proceed in the same fashion as above with the second-order linear differential equation

$$(4.6) \qquad L(y) \equiv a_0(x)y'' + a_1(x)y' + a_2(x)y = 0,$$

where we now assume that $a_0(x)$, $a_0'(x)$, $a_0''(x)$, $a_1(x)$, $a_1'(x)$, $a_2(x)$ are continuous and $a_0(x) \neq 0$ on some interval I. We then seek a function $z(x)$ such that

$$(4.7) \qquad z(x)[a_0(x)y'' + a_1(x)y' + a_2(x)y] \equiv \frac{d}{dx}[k(x)y' + m(x)y].$$

By the identity (4.7) is meant a formal identity in the quantities x, y, y', y''. The functions $k(x)$ and $m(x)$ are to be determined as well as $z(x)$. We have the following equations for the determination of these functions:

$$(4.8) \qquad \begin{aligned} a_0(x)z(x) &= k(x), \\ a_1(x)z(x) &= k'(x) + m(x), \\ a_2(x)z(x) &= m'(x). \end{aligned}$$

These equations lead at once to the differential equation

$$(4.9) \qquad M(z) \equiv [a_0(x)z]'' - [a_1(x)z]' + [a_2(x)z] = 0$$

for the determination of the function $z(x)$. The functions $k(x)$ and $m(x)$ are then readily determined from the first two of equations (4.8). The expression $M(z)$ is called the *adjoint* of $L(y)$.

When $L(y) \equiv M(y)$, the equation

$$(4.10) \qquad L(y) \equiv a_0(x)y'' + a_1(x)y' + a_2(x)y = 0$$

is said to be *self-adjoint*. It will be seen that the differential equation (4.10) is self-adjoint if and only if

$$(4.11) \qquad a_1(x) = a_0'(x).$$

When (4.11) holds, the differential equation (4.10) can be written in the form

$$(4.12) \qquad [r(x)y']' + p(x)y = 0,$$

where we have written $r(x) = a_0(x)$ and $p(x) = a_2(x)$.

If $r(x)$, $r'(x)$, and $p(x)$ are continuous and $r(x) > 0$ on an interval I, the equation (4.12) can be written in the form

$$r(x)y'' + r'(x)\dot{y}'(x) + p(x)y = 0,$$

where the coefficients are continuous and the coefficient of y'' is positive on I.

Conversely, every equation

(4.13) $a(x)y'' + b(x)y' + c(x)y = 0,$

where $a(x)$, $b(x)$, and $c(x)$ are continuous and $a(x) > 0$ on I, can be written in the form (4.12); that is, in self-adjoint form. We multiply both members of (4.13) by the function

$$\frac{1}{a(x)} e^{\int [b(x)/a(x)]\, dx}.$$

Then

$$r(x) = e^{\int [b(x)/a(x)]\, dx},$$

and

$$p(x) = \frac{c(x)}{a(x)} e^{\int [b(x)/a(x)]\, dx}.$$

Note that when $a_1(x) = 0$, equation (4.10) may be put in self-adjoint form simply by dividing through by $a_0(x)$.

Example.   To put the differential equation

$$x^2 y'' + xy' + y = 0 \qquad (x > 0)$$

in self-adjoint form, we divide through by x^2, obtaining

$$y'' + \frac{1}{x} y' + \frac{1}{x^2} y = 0.$$

Multiplying through by $e^{\int dx/x} = x$, we have

$$(xy')' + \frac{1}{x} y = 0,$$

and the equation is in self-adjoint form.

Example.   Find functions $z(x)$, $k(x)$, and $m(x)$ such that

$$z(x)[x^2 y'' - 2xy' + 2y] \equiv \frac{d}{dx}[k(x)y' + m(x)y].$$

The function $z(x)$ can be found by solving equation (4.9) with $a_0 = x^2$, $a_1 = -2x$, $a_2 = 2$.   We have

$$(x^2 z)'' + (2xz)' + (2z) = 0,$$

or

$$x^2 z'' + 6xz' + 6z = 0.$$

Linearly independent solutions of this equation are x^{-2} and x^{-3}. Accordingly, we may take

$$z(x) = \frac{1}{x^2},$$

$$k(x) = 1,$$

$$m(x) = -\frac{2}{x}$$

[choosing $z(x) = x^{-3}$, or any linear combination $c_1 x^{-2} + c_2 x^{-3}$, would similarly provide a solution to the problem]. This information would enable us to solve the differential equation

$$x^2 y'' - 2xy' + 2y = 0,$$

for an equivalent equation is then

$$\frac{d}{dx}\left[y' - \frac{2}{x} y \right] = 0.$$

It follows that

$$y' - \frac{2}{x} y = -c_1 \qquad (c_1 \text{ constant}).$$

An integrating factor for this first-order linear differential equation is readily found to be x^{-2}. Thus,

$$(x^{-2}y)' = -c_1 x^{-2},$$

and

$$y = c_1 x + c_2 x^2.$$

Unfortunately, the adjoint equation is ordinarily quite as difficult to solve as the original equation.

Reduction of the order of a differential equation. When a solution $y_1(x) \neq 0$ of a second-order linear differential equation

(4.14) $$a(x)y'' + b(x)y' + c(x)y = 0$$

is known, a second linearly independent solution may be found by means of the substitution

(4.15) $$y = y_1(x)v.$$

Before demonstrating this result we shall apply the method to an example.

Example.   It is readily verified that x is a solution of the differential equation

(4.16) $$y'' - xy' + y = 0.$$

We make the substitution

$$y = xv$$

obtaining

$$xv'' + (2 - x^2)v' = 0,$$

or,

$$xp' + (2 - x^2)p = 0 \qquad (p = v').$$

The last equation is a linear differential equation of first order in p. It has the solution

$$p = \frac{1}{x^2} e^{x^2/2}$$

on any interval not containing the origin. Accordingly, a second solution of (4.16) is

$$y_2(x) = x \int_1^x \frac{1}{x^2} e^{x^2/2} \, dx.$$

It is readily seen that x and $y_2(x)$ are linearly independent solutions of (4.16). Note that properly defined at $x = 0$, $y_2(x)$ must be of class C'' on $(-\infty, \infty)$ [Why? What value should be assigned to $y_2(0)$?].

We assume, as usual, that in (4.14) $a(x) > 0$ and $a(x)$, $b(x)$, $c(x)$ are continuous on $[a, b]$. The substitution $y = y_1(x)v$ yields

(4.17) $$ay_1v'' + (2ay_1' + by_1)v' + (ay_1'' + by_1' + cy_1)v = 0.$$

The coefficient of v is zero, for we have assumed that $y_1(x)$ is a solution of (4.14); accordingly, (4.17) may be written as

(4.18) $$ay_1p' + (2ay_1' + by)p = 0 \qquad (p = v').$$

Equation (4.18) is a first-order linear differential equation in the variable p. A solution is readily found to be

$$p = v' = \frac{1}{y_1^2} \exp\left[-\int_a^x \frac{b(x)}{a(x)} \, dx\right].$$

It follows that a second solution of (4.14) may be written in the form

$$(4.19) \qquad y_1(x) \int_a^x \frac{1}{y_1^2(x)} \left\{ \exp\left[-\int_a^x \frac{b(x)}{a(x)}\, dx \right] \right\} dx.$$

The difficulties that may exist at a zero of $y_1(x)$ will be dealt with below.

The substitution $y = y_1(x)v$ is also useful in solving nonhomogeneous differential equations

$$a(x)y'' + b(x)y' + c(x)y = f(x),$$

when one solution $y_1(x)$ of the corresponding homogeneous equation (4.14) is known.

An alternate (and simpler) method of finding a second solution of equation (4.14) is the following. Write (4.14) in self-adjoint form

$$(4.20) \qquad [r(x)y']' + p(x)y = 0,$$

where $r(x) > 0$ and $r(x)$ and $p(x)$ are continuous on $[a, b]$. If $y_1(x) \neq 0$ is a known solution of (4.20), there is a second linearly independent solution $y_2(x)$ that satisfies the differential equation

$$(4.21) \qquad r(x)[y_1(x)y_2' - y_1'(x)y_2] = 1 \qquad (Abel's\ formula,\ \text{p. 92});$$

that is, $y_2(x)$ is a solution of the first-order linear differential equation (4.21). This equation is readily solved, and it follows that $y_2(x)$ may be taken as

$$(4.22) \qquad y_2(x) = y_1(x) \int_a^x \frac{dx}{r(x)y_1^2(x)},$$

except possibly in the zeros of $y_1(x)$.

The difficulty in the zeros of $y_1(x)$ is more apparent than real, for if $y_1(x_0) = 0$, by l'Hospital's rule,

$$\lim_{x \to x_0} y_2(x) = \lim_{x \to x_0} \frac{\displaystyle\int_a^x \frac{dx}{r(x)y_1^2(x)}}{\dfrac{1}{y_1(x)}} = \lim_{x \to x_0} \frac{\dfrac{1}{r(x)y_1^2(x)}}{\dfrac{-y_1'(x)}{y_1^2(x)}} = \lim_{x \to x_0} \frac{-1}{r(x)y_1'(x)}$$

$$(4.23) \qquad = -\frac{1}{r(x_0)y_1'(x_0)}.$$

Note that $y_1'(x_0) \neq 0$. Thus, (4.22), with $y_2(x_0)$ defined by (4.23), provides a second solution of (4.20) on the entire interval $[a, b]$. That this solution is linearly independent of $y_1(x)$ may be seen in a number of ways. One way is to observe that the integral in (4.22) is not identically constant.

It will be observed that (4.22) is much simpler in form than (4.19). There is no advantage, however, in putting an equation like

(4.24) $$y'' - 3y' + 2y = 0$$

in self-adjoint form before making the substitution $y = y_1(x)v$. For example, if we are given that e^x is a solution of (4.24), we make the immediate substitution $y = e^x v$ in (4.24). There are, however, many practical situations, as well as theoretical ones, in which the use of the self-adjoint form is of distinct advantage. Some of these will appear later in this book.

Exercises

1. Put the following differential equations in self-adjoint form:

   (a) $xy'' - y' + x^2 y = 0$;
   (b) $k(x)y'' + m(x)y = 0$;
   (c) $y'' - 3y' + 2y = 0$;
   (d) $x^2 y'' + xy' + (x^2 - n^2)y = 0$;
   (e) $(1 - x^2)y'' - 2xy' + (n^2 + n)y = 0$.

2. Find a function $z(x)$ such that

   (a) $z(x)[y'' + y] \equiv \dfrac{d}{dx}[k(x)y' + m(x)y]$;

   (b) $z(x)[y'' - y] \equiv \dfrac{d}{dx}[k(x)y' + m(x)y]$;

   (c) $z(x)[y'' + 3y' + 2y] \equiv \dfrac{d}{dx}[k(x)y' + m(x)y]$.

3. Given that one solution of $y'' + a^2 y = 0$ is $\sin ax$, find a second linearly independent solution, using the method of the text.

4. Given that one solution of $y'' - 4y' + 4y = 0$ is e^{2x}, use the method of the text to find a second linearly independent solution.

5. Given that one solution of $y'' - 2ay' + a^2 y = 0$ is e^{ax} $(a \neq 0)$, use the method of the text to find a second linearly independent solution.

6. Given that $x + 1$ is a solution of $y'' - 2(x + 1)y' + 2y = 0$, find the general solution.

7. Find the general solution of the differential equation $x^3 y'' - xy' + y = 0$ $(x > 0)$. (*Hint.* Guess one solution.)

8. Use Exercise 7 to solve the differential equation $x^4 y'' - x^2 y' + xy = 1$ $(x > 0)$.

9.   One solution of $x^2 y'' + xy' + (x^2 - \frac{1}{4})y = 0$ is $x^{-1/2} \sin x$. Find the general solution.

10.   Show that except for a multiplicative factor (4.19) reduces to (4.22) when $a_0(x) = r(x)$ and $a_1(x) = r'(x)$.

11.   Show that the solution (4.22) is linearly independent of the given solution $y_1(x)$.

Answers

1.   (a) $[(1/x)y']' + y = 0$;
   (c) $(e^{-3x}y')' + 2e^{-3x}y = 0$;
   (g) $(x^b y')' + cx^{b-2}y = 0$.

2.   (a) $z(x) = \sin x$,
   $k(x) = \sin x$,
   $m(x) = -\cos x$.

6. $c_1(x + 1) + c_2(x + 1) \int_0^x \dfrac{1}{(x + 1)^2} e^{(x+1)^2}\, dx$.

8.   A particular solution is $-1 + \frac{1}{2}x^{-1}$.

9. $c_1 \dfrac{\sin x}{\sqrt{x}} + c_2 \dfrac{\cos x}{\sqrt{x}}$.

5   The Riccati equation

If the substitution $z = \dfrac{r(x)y'}{y}$ is made in the self-adjoint differential equation

$$(5.1) \qquad\qquad [r(x)y']' + p(x)y = 0,$$

where $r(x)$ and $p(x)$ are continuous on an interval $a \le x \le b$, we obtain

$$(yz)' + p(x)y = 0,$$

or

$$(5.2) \qquad\qquad z' + \frac{1}{r(x)} z^2 + p(x) = 0.$$

Equation (5.2) is a *Riccati equation*. The general Riccati equation is usually written as

(5.3) $$z' + a(x)z + b(x)z^2 + c(x) = 0,$$

where we shall suppose $a(x)$, $b(x)$, and $c(x)$ are continuous on the interval $a \leq x \leq b$. Equation (5.3) is only apparently more general than equation (5.2), since the substitution in (5.3) of

(5.4) $$w = e^{\int_a^x a(x)\,dx} z$$

reduces this equation to

(5.5) $$w' + q(x)w^2 + p(x) = 0,$$

where

$$q(x) = b(x)e^{-\int_a^x a(x)\,dx},$$

$$p(x) = c(x)e^{\int_a^x a(x)\,dx}.$$

If $b(x) \equiv 0$, equation (5.3) is, of course, linear and it is immediately integrable. If $b(x) \not\equiv 0$ on any subinterval of $[a, b]$, to study the solutions of (5.3) we may employ the substitution (5.4) to reduce (5.3) to the form (5.5). The substitution $qw = \dfrac{y'}{y}$ then reduces (5.5) to the form (5.1), where $r(x) = \dfrac{1}{q(x)}$. The zeros of $q(x)$ are then singular points of the differential equation (5.1). It will be observed that these successive substitutions may be replaced by the substitution $bz = \dfrac{y'}{y}$.

Example.   Study the solutions of the Riccati equation

(5.6) $$w' - w^2 - 1 = 0.$$

This equation is already in the form (5.5), where $q(x) = -1$ and $p(x) = -1$. The substitution $-w = \dfrac{y'}{y}$ leads then to the linear self-adjoint differential equation

$$y'' + y = 0,$$

the general solution of which is $c_1 \sin x + c_2 \cos x$. The null solution $(c_1 = c_2 = 0)$ leads to no solution w. All other solutions y provide solutions

(5.7) $$w = \frac{c_1 \cos x - c_2 \sin x}{c_1 \sin x + c_2 \cos x} \qquad (c_1 \text{ and } c_2 \text{ not both zero})$$

of (5.6). The choice $c_1 = 0$ leads to the particular solution $w = -\tan x$.

Example.   Study the solutions of the Riccati equation

$$(5.8) \qquad\qquad z' + z - e^x z^2 - e^{-x} = 0.$$

The substitution $w = e^x z$ reduces this equation to (5.6). Thus solutions z of (5.8) are

$$(5.9) \qquad z = -e^{-x}\frac{c_1 \cos x - c_2 \sin x}{c_1 \sin x + c_2 \cos x} \qquad (c_1 \text{ and } c_2 \text{ not both zero}).$$

The student will have observed that the method of the text enables us to find an infinity of solutions of a Riccati equation. Two observations come to mind. First, may there not be solutions of the Riccati equation other than those obtained in this way? And second, we note that the solutions so obtained [see (5.7) and (5.9)] contain *two* arbitrary constants, c_1 and c_2. The Riccati equation is a first-order differential equation, and our experience suggests that *one* arbitrary constant might be expected.

Let us deal with the latter question first. In general, following the method of the text, we shall be led to solutions z of (5.3) of the form

$$(5.10) \qquad\qquad z = -f(x)\frac{c_1 u'(x) + c_2 v'(x)}{c_1 u(x) + c_2 v(x)},$$

where $f(x) \neq 0$, c_1 and c_2 are not both zero, and $u(x)$ and $v(x)$ are linearly independent solutions of the related linear differential equation (5.1). The solution $z = \dfrac{-f(x)v'(x)}{v(x)}$ is obtained by setting $c_1 = 0$ (hence $c_2 \neq 0$). For all other solutions given by (5.10), $c_1 \neq 0$, and (5.10) may accordingly be written in the form

$$(5.11) \qquad\qquad z = -f(x)\frac{u'(x) + kv'(x)}{u(x) + kv(x)},$$

where $k = \dfrac{c_2}{c_1}$ is an arbitrary constant. Thus the solutions which we have depend essentially on one arbitrary constant. The exceptional solution $\dfrac{-f(x)v'(x)}{v(x)}$ may be regarded (and commonly is) as being obtained from (5.11) when k is set equal to infinity. The student will no doubt prefer the form (5.10) which, as we shall see, yields *all* solutions.

To prove the last remark, note first that if $z(x)$ is a solution of (5.3), $w(x)$, defined by (5.4), is a solution of (5.5), and conversely. We have then the following result.

Theorem 5.1. If $q(x) \neq 0$, and $p(x)$ and $q(x)$ are continuous on the interval $a \leq x \leq b$, every solution $w(x)$ of (5.5) may be written in the form

(5.12)
$$\frac{1}{q(x)} \frac{c_1 u'(x) + c_2 v'(x)}{c_1 u(x) + c_2 v(x)},$$

where c_1 and c_2 are constants, not both zero, and where $u(x)$ and $v(x)$ are linearly independent solutions of the differential equation

(5.13)
$$\left[\frac{1}{q(x)} y' \right]' + p(x)y = 0.$$

Conversely, if $u(x)$ and $v(x)$ are linearly independent solutions of (5.13) and if c_1 and c_2 are any constants, not both zero, the function (5.12) is a solution of (5.5) on any interval in which $c_1 u(x) + c_2 v(x) \neq 0$.

To prove the theorem, let $w(x)$ be an arbitrary solution of (5.5). We may then readily verify that the function

$$e^{\int_a^x q(x)w(x)\, dx}$$

is a solution not equal to zero of (5.13). Thus there exist constants not both zero such that

$$e^{\int_a^x q(x)w(x)\, dx} \equiv c_1 u(x) + c_2 v(x).$$

Upon differentiating both members of this identity the first statement of the theorem follows at once.

Conversely, we assume $u(x)$ and $v(x)$ are linearly independent solutions of (5.13), and $L(x) \equiv c_1 u(x) + c_2 v(x) \neq 0$ on $\alpha \leq x \leq \beta$. It is then easily seen that $\dfrac{L'(x)}{q(x)L(x)}$ is a solution of (5.5). For,

$$\left[\frac{L'(x)}{q(x)L(x)} \right]' + q(x)\left[\frac{L'(x)}{q(x)L(x)} \right]^2 + p(x)$$

$$\equiv \left[\frac{L'(x)}{q(x)} \right]' \frac{1}{L(x)} - \frac{[L'(x)]^2}{q(x)L^2(x)} + q(x)\left[\frac{L'(x)}{q(x)L(x)} \right]^2 + p(x) \equiv 0.$$

The proof of the theorem is complete.

Exercises

1.   Show that if z, z_1, z_2, z_3 are any four different solutions of the Riccati equation (5.3), then

$$\frac{z - z_2}{z - z_1} \frac{z_3 - z_1}{z_3 - z_2} \equiv \text{constant.}$$

[*Hint.* First show that $u = \dfrac{1}{z - z_1}$ is a solution of the linear differential

equation $u' - (a + 2bz_1)u = b$. Similarly, $u_1 = \dfrac{1}{z_2 - z_1}$ and $u_2 = \dfrac{1}{z_3 - z_1}$
are solutions of this differential equation. Consider the differences $u - u_1$
and $u_2 - u_1$.]

2. Find all solutions of the following Riccati equations:
  (a) $z' + z^2 - 1 = 0$;
  (b) $z' + z^2 - z - 2 = 0$;
  (c) $z' + z^2 - 2z + 2 = 0$;
  (d) $x^2 z' - 2xz + x^2 z^2 + 2 = 0$;
  (e) $x^2 z' - 3xz + z^2 + 2x^2 = 0$;
  (f) $z' + z^2 + a^2 x^{-2} = 0$ (a constant);
  (g) $xz' - z + xz^2 + x^3 = 0$.

3. Find four distinct solutions for each of the differential equations in Exercise
2, and verify the conclusion of Exercise 1.

Answers

2. (a) $\dfrac{c_1 e^x - c_2 e^{-x}}{c_1 e^x + c_2 e^{-x}}$;   (b) $\dfrac{2c_1 e^{2x} - c_2 e^{-x}}{c_1 e^{2x} + c_2 e^{-x}}$;

(c) $\dfrac{(c_1 - c_2)\sin x + (c_1 + c_2)\cos x}{c_1 \sin x + c_2 \cos x}$;   (d) $\dfrac{c_1 + 2c_2 x}{c_1 x + c_2 x^2}$;

(e) $\dfrac{(c_1 - c_2)x \sin \log |x| + (c_1 + c_2)x \cos \log |x|}{c_1 \sin \log |x| + c_2 \cos \log |x|}$;

(g) $\dfrac{x[c_1 \cos (x^2/2) - c_2 \sin (x^2/2)]}{c_1 \sin (x^2/2) + c_2 \cos (x^2/2)}$.

6   Green's function

We return to equation (3.1), which we rewrite for convenience:

(6.1) $a_0(x)y'' + a_1(x)y' + a_2(x)y = f(x).$

Here, $a_0(x)$, $a_1(x)$, $a_2(x)$, and $f(x)$ are continuous and $a_0(x) \neq 0$ on an interval
$[a, b]$. The corresponding homogeneous equation is

(6.2) $a_0(x)y'' + a_1(x)y' + a_2(x)y = 0.$

Define the function

$$(6.3) \quad G(x, t) = -\frac{1}{a_0(t)} \frac{\begin{vmatrix} y_1(x) & y_2(x) \\ y_1(t) & y_2(t) \end{vmatrix}}{\begin{vmatrix} y_1(t) & y_2(t) \\ y_1'(t) & y_2'(t) \end{vmatrix}} = -\frac{1}{a_0(t)w(t)} \begin{vmatrix} y_1(x) & y_2(x) \\ y_1(t) & y_2(t) \end{vmatrix}$$

$$(x_0 \leq t \leq x),$$

where $y_1(x)$ and $y_2(x)$ are linearly independent solutions of equation (6.2). It follows from (3.7) and Theorem 3.1 that a particular solution of (6.1) is

$$\int_{x_0}^{x} G(x, t) f(t) \, dt$$

and that the general solution of (5.1) may be written as

$$(6.4) \qquad\qquad c_1 y_1(x) + c_2 y_2(x) + \int_{x_0}^{x} G(x, t) f(t) \, dt.$$

The function $G(x, t)$ defined in (6.3) is known as *Green's function* for the homogeneous equation (6.2). Clearly, if two linearly independent solutions of the homogeneous equation (6.2) are known, $G(x, t)$ can be constructed from (6.3), and then (6.4) provides the general solution of the nonhomogeneous equation (6.1) for any function $f(x)$ that is continuous on $[a, b]$. We note that when $x = t$, G vanishes; that is,

$$G(t, t) = 0.$$

As an illustration of the procedure, consider the homogeneous equation

$$y'' - 3y' + 2y = 0 \qquad (-\infty < x < \infty),$$

linearly independent solutions of which are

$$e^x, \qquad e^{2x}.$$

Then,

$$a_0(t) = 1, \qquad w(t) = \begin{vmatrix} e^t & e^{2t} \\ e^t & 2e^{2t} \end{vmatrix} = e^{3t},$$

and

$$G(x, t) = -e^{-3t} \begin{vmatrix} e^x & e^{2x} \\ e^t & e^{2t} \end{vmatrix} = e^{2(x-t)} - e^{(x-t)} \qquad (x_0 \leq t \leq x).$$

Accordingly, if $f(x)$ is any continuous function, a particular solution of

(6.5) $$y'' - 3y' + 2y = f(x)$$

may be written in the form

(6.6) $$\int_0^x [e^{2(x-t)} - e^{(x-t)}]f(t)\,dt,$$

where we have taken $x_0 = 0$. The general solution of (6.5) can then be written as

(6.7) $$c_1 e^x + c_2 e^{2x} + \int_0^x [e^{2(x-t)} - e^{(x-t)}]f(t)\,dt.$$

If, for example, $f(x) = x$, (6.6) becomes

$$\int_0^x [e^{2(x-t)} - e^{(x-t)}]t\,dt = -e^x + \frac{1}{4}e^{2x} + \frac{x}{2} + \frac{3}{4},$$

and the general solution (6.7) can be written as

$$c_1 e^x + c_2 e^{2x} - e^x + \frac{1}{4}e^{2x} + \frac{x}{2} + \frac{3}{4},$$

or

$$k_1 e^x + k_2 e^{2x} + \frac{x}{2} + \frac{3}{4},$$

where $k_1 = c_1 - 1$ and $k_2 = c_2 + \frac{1}{4}$.

The particular solution

(6.8) $$y_0(x) = \int_{x_0}^x G(x, t)f(t)\,dt$$

of equation (6.1) has the property that

$$y_0(x_0) = 0,$$

since

$$\int_{x_0}^{x_0} G(x_0, t)f(t)\,dt = 0.$$

Further,

$$y_0'(x) = G(x, x)f(x) + \int_{x_0}^x G_x(x, t)f(t)\, dt,$$

and, hence,

$$y_0'(x_0) = G(x_0, x_0)f(x_0)$$
$$= 0.$$

Thus, (6.8) provides the (unique) particular solution $y_0(x)$ of (6.1) that has the property that

$$y_0(x_0) = y_0'(x_0) = 0.$$

We next write
$$L(y) = a_0(x)y'' + a_1(x)y' + a_2(x)y,$$

where we now assume that $a_0(x)$, $a_0'(x)$, $a_0''(x)$, $a_1(x)$, $a_1'(x)$, $a_2(x)$ are continuous and $a_0(x) \neq 0$ on some interval $[a, b]$. We observed in Sec. 4 that the adjoint $M(z)$ of $L(y)$ was given by

$$M(z) = [a_0(x)z]'' - [a_1(x)z]' + [a_2(x)z].$$

Theorem 6.1. (LA GRANGE)

(6.9) $$vL(u) - uM(v) \equiv \frac{d}{dx}P(u, v),$$

where

$$P(u, v) = u[a_1v - (a_0v)'] + u'(a_0v).$$

The identity (6.9) is a formal identity in u, v, u', v', u'', v'', the proof of which is left to the student. It is also valid when

(6.10) $$L(y) = a_0y^{(n)} + a_1y^{(n-1)} + \cdots + a_{n-1}y' + a_ny.$$

In this case,

$$M(z) = (-1)^n(a_0z)^{(n)} + (-1)^{n-1}(a_1z)^{(n-1)} + \cdots - (a_{n-1}z)' + a_nz,$$

and it can be shown that

$$P(u, v) = u[a_{n-1}v - (a_{n-2}v)' + \cdots + (-1)^{n-1}(a_0v)^{(n-1)}]$$
$$+ u'[a_{n-2}v - (a_{n-3}v)' + \cdots + (-1)^{n-2}(a_0v)^{n-2)}]$$
$$+ \cdots + u^{(n-1)}a_0v.$$

One assumes whatever differentiability of the functions $a_i(x)$ that one requires; it is generally sufficient to assume that $a_i(x)$ is of class C^{n-i} $(i = 0, 1, \ldots, n)$ on some interval $[a, b]$.

When the identity (6.9) is integrated, the result

$$\int_a^b [vL(u) - uM(v)]\, dx = [P(u, v)]_a^b$$

is known as *Green's formula*.

Exercises

1. Prove Theorem 6.1.

2. Given the differential equation

(1) $\qquad\qquad y'' - 4y' + 3y = f(x) \qquad (-\infty < x < \infty)$

find two linearly independent solutions of

$$y'' - 4y' + 3y = 0.$$

Compute Green's function $G(x, t)$ and then compute the particular solution

$$\int_0^x G(x, t)f(t)\, dt$$

of (1) when (a) $f(x) = 3$; (b) $f(x) = x$. Check your answers. What is the general solution of (1) in each case?

3. Do the same as Exercise 2 when the differential equation (1) is

(a) $y'' + y = f(x) \qquad (-\infty < x < \infty)$;
(b) $y'' + 4y' + 4y = f(x) \qquad (-\infty < x < \infty)$;
(c) $x^2 y'' - 2xy' + 2y = f(x) \qquad (1 \le x < \infty)$;
(d) $4x^2 y'' + y = f(x) \qquad (1 \le x < \infty)$.

4. Show that $G(x, t)$ is, for each t, a solution of equation (6.2).

5. Show that

$$G_x(t, t) \equiv \frac{\partial}{\partial x} G(x, t)\Big|^{x=t} = \frac{1}{a_0(t)}.$$

5

Oscillation theory for linear differential equations of second order

1 Self-adjoint linear differential equations of second order

The importance of the self-adjoint form in the study of linear differential equations of second order can hardly be overemphasized. It arises naturally in mechanics; it has a central role in the calculus of variations. The student will observe its use throughout the present chapter as we study the behavior of solutions of the linear differential equation of second order.

A self-adjoint linear differential equation of second order, it will be recalled, is a differential equation of the form

$$(1.1) \qquad [r(x)y']' + p(x)y = 0,$$

where $r(x)$ and $p(x)$ are continuous and $r(x) > 0$ on an interval I. The results will apply to the differential equation

$$(1.2) \qquad a(x)y'' + b(x)y' + c(x)y = 0,$$

where $a(x) > 0$ and $a(x)$, $b(x)$, and $c(x)$ are continuous on the interval I because, as we have seen, equation (1.2) can be put in self-adjoint form by multiplying both members of the equation by the function

$$\frac{1}{a(x)} e^{\int_{x_0}^{x} [b(x)/a(x)]\, dx}$$

Then

$$r(x) = e^{\int_{x_0}^{x} [b(x)/a(x)]}\, dx, \qquad p(x) = \frac{c(x)}{a(x)}\, r(x).$$

2   Abel's formula

Let $u(x)$ and $v(x)$ be any two solutions of (1.1). We shall show that

$$(2.1) \qquad\qquad r(x)[u(x)v'(x) - u'(x)v(x)] \equiv k,$$

a constant. Since $u(x)$ and $v(x)$ are solutions of (1.1), we have

$$[r(x)u'(x)]' + p(x)u(x) \equiv 0,$$
$$[r(x)v'(x)]' + p(x)v(x) \equiv 0.$$

If we multiply both sides of the first identity by $-v(x)$, both members of the second identity by $u(x)$, and add, we obtain

$$u(x)[r(x)v'(x)]' - v(x)[r(x)u'(x)]' \equiv 0.$$

We integrate both members of the last identity from a to x, and integration by parts yields

$$r(x)[u(x)v'(x) - u'(x)v(x)] \equiv r(a)[u(a)v'(a) - u'(a)v(a)];$$

that is to say, the left-hand member is constant. As we saw in Chapter 4, the quantity in brackets $u(x)v'(x) - u'(x)v(x)$ is the wronskian of the solutions $u(x)$ and $v(x)$. Thus the constant k is zero if and only if $u(x)$ and $v(x)$ are linearly dependent. The identity (2.1) is known as *Abel's formula*.

3   The number of zeros on a finite interval

In this section we shall prove the following result.

Theorem 3.1 *If $r(x) > 0$ and $r(x)$ and $p(x)$ are continuous on the interval $a \leq x \leq b$, the only solution of equation* (1.1) *which vanishes infinitely often on this interval is the null solution.*

This theorem may be proved as follows. Suppose the solution $y(x)$ has an infinity of zeros on this interval. The set of zeros will then have a limit point x^* on the interval, and there will exist a sequence $\{x_n\}_0^\infty$ of the zeros converging to x^* with

$$x_n \neq x^* \qquad (n = 0, 1, 2, \ldots).$$

We shall show that $y(x^*) = y'(x^*) = 0$. It will follow from the corollary to Theorem 1.1, Chapter 4, that $y(x) \equiv 0$. To that end note that $y(x)$ is a continuous function of x. Thus,

$$\lim_{x \to x^*} y(x) = y(x^*)$$

as x tends to x^* through any sequence of numbers on $[a, b]$. Choosing as this

sequence the numbers x_0, x_1, $x_2 \ldots$, we see that $y(x^*) = 0$.
Next, note that

$$\lim_{x \to x^*} \frac{y(x) - y(x^*)}{x - x^*} = y'(x^*).$$

Since $y'(x^*)$ is known to exist, we may evaluate the limit by letting x tend to x^* through members of the sequence x_0, x_1, $x_2 \ldots$. Thus $y'(x^*) = 0$, and the proof is complete.

Definition. If the differential equation (1.1) *has a nonnull solution that vanishes infinitely often on* I, *the solutions of* (1.1) *are said to be oscillatory on* I.

Thus, solutions of

$$(xy')' + \frac{1}{x}y = 0$$

are oscillatory both on $(0, 1]$ and on $[1, \infty)$.

4 The Sturm separation theorem

We are now prepared to prove a fundamental result due to Sturm. It will be helpful in that proof to have available the following lemma.

Lemma. If two solutions $u(x)$ *and* $v(x)$ *of* (1.1) *have a common zero, they are linearly dependent. Conversely, if* $u(x)$ *and* $v(x)$ *are linearly dependent solutions, neither identically zero, then if one of them vanishes at* $x = x_0$, *so does the other.*

To prove the first statement of the lemma, we employ Abel's formula

(4.1) $$r(x)[u(x)v'(x) - u'(x)v(x)] \equiv k.$$

Let the common zero of $u(x)$ and $v(x)$ be $x = x_0$, and replace x by x_0 in (4.1). It follows that the constant k is zero, and hence that $u(x)$ and $v(x)$ are linearly dependent.

We could also employ (4.1) to prove the second statement of the lemma. It will be somewhat simpler, however, to use the definition of linear dependence which states that there exist constants c_1 and c_2 not both zero such that

$$c_1 u(x) + c_2 v(x) \equiv 0.$$

Since we have assumed that neither $u(x)$ nor $v(x)$ is identically zero, we see that both c_1 and c_2 are different from zero. Thus, if $u(x_0) = 0$, then $v(x_0) = 0$ also.

The proof of the lemma is complete.

Theorem 4.1. THE STURM SEPARATION THEOREM. *If $u(x)$ and $v(x)$ are linearly independent solutions of* (1.1), *between two consecutive zeros of $u(x)$ there will be precisely one zero of $v(x)$.*

It should be noted, of course, that not all equations (1.1) have a nonnull solution that vanishes twice on I.

Let $x = x_0$ and $x = x_1$ be two consecutive zeros of $u(x)$ (see Fig. 5.1), and let x_0 be the smaller of the two. We may without loss in generality assume that $u(x) > 0$ for $x_0 < x < x_1$. This follows from the fact that $-u(x)$ is also a solution having the same zeros as $u(x)$. It follows that $u'(x_0) > 0$ and $u'(x_1) < 0$. Similarly, we may suppose without loss in generality that $v(x_0) > 0$, since $v(x_0) \neq 0$ by the lemma.

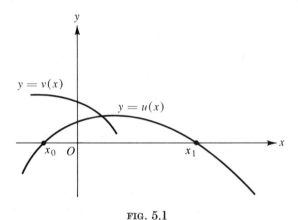

FIG. 5.1

To prove the theorem, we set $x = x_0$ in equation (4.1), and note that therefore $k < 0$. It follows that

$$r(x_1)[u(x_1)v'(x_1) - u'(x_1)v(x_1)] < 0.$$

But $u(x_1) = 0$, $u'(x_1) < 0$; hence $v(x_1) < 0$. Since $v(x)$ is a continuous function, it must have at least one zero between x_0, where $v(x)$ is positive, and x_1, where $v(x)$ is negative. There cannot be more than one zero because the argument above, with the roles of $u(x)$ and $v(x)$ reversed, shows that between two consecutive zeros of $v(x)$ there must be at least one zero of $u(x)$.

The proof of the Sturm separation theorem is complete.

More generally, we have the following result.

Theorem 4.2. *Let $u(x)$ and $v(x)$ be functions of class C' on $[a, b]$ and suppose that $v(a) = v(b) = 0$, $v(x) \neq 0$ on (a, b). If*

$$w(x) = u(x)v'(x) - v(x)u'(x) \neq 0$$

on $[a, b]$, $u(x)$ vanishes precisely once on (a, b).

There will be no loss in generality in supposing that $w(x) > 0$ on $[a, b]$ and that $v(x) > 0$ on (a, b). Then,

$$w(a) = u(a)v'(a) > 0 \qquad \text{and} \qquad w(b) = u(b)v'(b) > 0.$$

Because $v'(a) > 0$ and $v'(b) < 0$, it follows that $u(a) > 0$ and $u(b) < 0$. Accordingly, $u(x)$ must vanish at least once on (a, b). Note that inasmuch as $w(x) \neq 0$ on (a, b), a zero of $u(x)$ is necessarily simple. Suppose $u(x)$ had a second zero on (a, b). The same argument, with the roles of $u(x)$ and $v(x)$ interchanged, shows that $v(x)$ would have a zero between the two zeros of $u(x)$, contrary to hypothesis.

The proof of the theorem is complete.

The following corollary is an immediate consequence of the theorem.

Corollary. If $v(x)$ has an infinity of zeros and if $w(x) \neq 0$ on $[a, b)$, then
(a) $u(x)$ has an infinity of zeros on $[a, b)$ also;
(b) the zeros of both $u(x)$ and $v(x)$ are simple;
(c) the zeros of $u(x)$ and $v(x)$ separate each other.
Finally, b may be finite or infinite.

Note that if $b = \infty$, and if $w(x)$ is eventually (for x sufficiently large) of one sign, the conclusions of the corollary are valid for x sufficiently large.

A fundamental lemma. Although we shall not employ it here, the following result is of fundamental importance in many oscillation studies.† It is complementary to Theorem 4.2 and is included for completeness.

Lemma. Let $u(x)$ and $v(x)$ be functions of class C' on $[a, b]$ and suppose that $u(a) = u(b) = 0$ and that $v(x) \neq 0$ on $[a, b]$. There then exist constants c_1 and c_2, not both zero, such that the function $c_1 u(x) + c_2 v(x)$ has a double zero on (a, b).

To prove the lemma, consider the function

$$w(x) = u(x)v'(x) - u'(x)v(x).$$

---

† See, for example, W. Leighton and Z. Nehari, "On the oscillation of solutions of self-adjoint linear differential equations of fourth order, *Trans. Amer. Math. Soc.*, Vol. 89 (1958), pp. 325–377.

Suppose, without loss in generality, that $v(x) > 0$ on $[a, b]$, that $x = a$ and $x = b$ are consecutive zeros of $u(x)$, and that $u(x) > 0$ on (a, b). If the zeros of $u(x)$ are simple, $u'(a) > 0$, $u'(b) < 0$, and

$$w(a) = -u'(a)v(a) < 0, \qquad w(b) = -u'(b)v(b) > 0.$$

Thus $w(x)$ must vanish at some point $x = x_0$ on (a, b); that is,

$$\begin{vmatrix} u(x_0) & v(x_0) \\ u'(x_0) & v'(x_0) \end{vmatrix} = 0.$$

It follows that there exist constants c_1 and c_2 not both zero such that

(4.2)
$$c_1 u(x_0) + c_2 v(x_0) = 0,$$
$$c_1 u'(x_0) + c_2 v'(x_0) = 0.$$

That is, the function $c_1 u(x) + c_2 v(x)$ has a double zero at $x = x_0$.

If the zeros of $u(x)$ are not necessarily simple, the proof must be altered somewhat (it is also more general). In this situation, again assume without loss in generality that $x = a$ and $x = b$ are consecutive zeros of $u(x)$, with $u(x) > 0$ on (a, b), and that $v(x) > 0$ on $[a, b]$. We can then choose a constant c positive, but small enough that the curve $y = cv(x)$ crosses the curve $y = u(x)$ in at least two distinct points. Let $x = \alpha$ at the smallest such point and $x = \beta$ at the largest. Then,

$$cv'(\alpha) < u'(\alpha) \qquad \text{and} \qquad cv'(\beta) > u'(\beta).$$

This is clear geometrically and is readily argued analytically.

Consider the function

$$w(x) = u(x)cv'(x) - cv(x)u'(x)$$

and note that $u(\alpha) = cv(\alpha)$, $u(\beta) = cv(\beta)$. Then,

$$w(\alpha) = u(\alpha)[cv'(\alpha) - u'(\alpha)] < 0,$$
$$w(\beta) = u(\beta)[cv'(\beta) - u'(\beta)] > 0.$$

It follows that $w(x)$ has a zero $x = x_0$ on (α, β), and, consequently, that

$$u(x_0)v'(x_0) - v(x_0)u'(x_0) = 0 \qquad (a < \alpha < x_0 < \beta < b).$$

The remainder of the proof is as before, and the proof of the lemma is complete.

Note from (4.2) that if c_1 were zero, then $c_2 \neq 0$, $v(x_0) = 0$, contrary to hypothesis. Accordingly, we may conclude that $c_1 \neq 0$ and the conclusion of the theorem may be given as *there exists a constant k such that the function*

$u(x) - kv(x)$ has a double zero on (a, b), for k may be taken as $-c_2/c_1$. Geometrically, the lemma may be interpreted to mean that there exists a constant k with the property that the curves $y = kv(x)$ and $y = u(x)$ are tangent at a point $x = x_0$ of (a, b).

Exercises

1. Put the following differential equations in self-adjoint form:

    (a) $x^2 y'' + xy' + (x^2 - n^2)y = 0$ (n constant);
    (b) $(1 - x^2)y'' - 2xy' + n(n + 1)y = 0$ (n constant);
    (c) $(1 - x^2)y'' - xy' + n^2 y = 0$ (n constant);
    (d) $xy'' + (1 - x)y' + ny = 0$ (n constant);
    (e) $y'' - 2xy' + 2ny = 0$ (n constant);
    (f) $(1 - x^2)y'' - [(\alpha - \beta) + (\alpha + \beta)x]y' + n(n + \alpha + \beta + 1)y = 0$
       ($\alpha > -1, \beta > -1$).

2. Prove that between every pair of consecutive zeros of $\sin x$ there is one zero of $\sin x + \cos x$. (*Hint.* Use Sturm's theorem.)

3. If $u(x)$ is a solution of (1.1) such that $u(x_0) = u(x_1) = 0$ and $u(x) > 0$ ($x_0 < x < x_1$), prove that $u'(x_0) > 0$ and that $u'(x_1) < 0$.

4. Construct an example of a differential equation in the form (1.1) such that no nonnull solution has more than one zero [thus, there are equations (1.1) to which Sturm's theorem does not apply].

5. Show that between every pair of consecutive zeros of $\sin \log x$ there is a zero of $\cos \log x$.

6. Find a self-adjoint differential equation which has the solution $r(x)y'(x)$, where $y(x)$ is any solution of (1.1).

7. Prove that if in equation (1.1)

$$r(x)p(x) \equiv 1,$$

linearly independent solutions of (1.1) are $\sin v(x)$ and $\cos v(x)$, where

$$v(x) = \int p(x)\, dx.$$

What are linearly independent solutions of (1.1) when $r(x)p(x) \equiv k^2$ (k a positive constant)?

8. Let $m(x)$ be a function of class C' on the interval $a \leq x \leq b$, and suppose that $m(x) \neq 0$ on any subinterval of $[a, b]$. The differential equation

$$\left[\frac{1}{m(x)} y'\right]' + k^2 m(x)y = 0 \qquad (k \text{ constant}, \neq 0)$$

will then have a singular point whenever $m(x) = 0$. Prove, however, that all solutions of this equation are of class C'' on $[a, b]$.

9.   Given equation (1.1) with the usual conditions on the coefficients $r(x)$ and $p(x)$, discuss the possibility of solutions $u(x)$ and $v(x)$ having the configuration in Fig. 5.2.

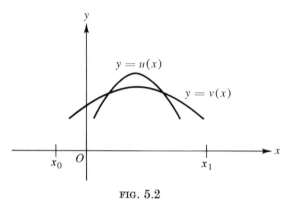

FIG. 5.2

Answer

6. $[(1/p)z']' + (1/r)z = 0$.

5   The Sturm comparison theorem

In this section we continue the study of solutions of differential equations of the form

(5.1) $[r(x)y']' + p(x)y = 0,$

where $r(x) > 0$, and $r(x)$ and $p(x)$ are continuous on the closed interval $a \leq x \leq b$. The student will recall that the Sturm separation theorem asserts that between two consecutive zeros of a solution of (5.1) there appears one zero of every linearly independent solution. Thus, speaking roughly, the number of zeros on an interval of any solution of (5.1) is about the same as the number any other solution possesses.

On the other hand, it is easy to see that solutions (for example, $\sin 2x$) of

$$y'' + 4y = 0$$

oscillate more frequently (that is, have more zeros) on the interval $0 \leq x \leq 2\pi$ than do the solutions of

$$y'' + y = 0$$

on that interval. A typical solution of the last equation is, of course, $\sin x$. The Sturm comparison theorem compares the rates of oscillation of solutions of two equations,

(5.2) $$[r(x)y']' + p(x)y = 0,$$

(5.3) $$[r(x)z']' + p_1(x)z = 0,$$

where $r(x) > 0$, $r(x)$, $p(x)$, $p_1(x)$ are continuous on $a \leq x \leq b$.

Theorem 5.1. THE STURM COMPARISON THEOREM. *If a solution $y(x)$ of (5.2) has consecutive zeros at $x = x_0$ and $x = x_1$ ($x_0 < x_1$), and if $p_1(x) \geq p(x)$ with strict inequality holding for at least one point of the closed interval $[x_0, x_1]$, a solution $z(x)$ of (5.3) which vanishes at $x = x_0$ will vanish again on the interval $x_0 < x < x_1$.*

That is to say, speaking roughly, the larger is $p(x)$, the more rapidly the solutions of (5.1) oscillate (see Fig. 5.3).

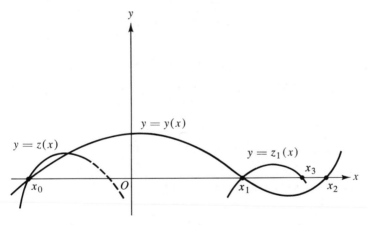

FIG. 5.3

To prove the theorem, we may suppose without loss in generality that $y(x) > 0$ on the interval $x_0 < x < x_1$, and that $y'(x_0) > 0$, $y'(x_1) < 0$, and $z'(x_0) > 0$. Since $y(x)$ and $z(x)$ are solutions, respectively, of equations (5.2) and (5.3), we have the identities

$$[r(x)y'(x)]' + p(x)y(x) \equiv 0,$$
$$[r(x)z'(x)]' + p_1(x)z(x) \equiv 0.$$

If we multiply the first of these by $-z(x)$ and the second by $y(x)$ and add, we may integrate both members of the resulting identity over the interval

$x_0 \leq x \leq x_1$, obtaining

$$r(x)[y(x)z'(x) - y'(x)z(x)]\Big|_{x_0}^{x_1} + \int_{x_0}^{x_1} [p_1(x) - p(x)]y(x)z(x)\, dx = 0,$$

or

(5.4) $$r(x_1)y'(x_1)z(x_1) = \int_{x_0}^{x_1} [p_1(x) - p(x)]y(x)z(x)\, dx.$$

Now suppose $z(x) > 0$ on $x_0 < x < x_1$. Then the integral in (5.4) is positive while the left-hand member is not. From this contradiction we infer the truth of the theorem.

We may note further that if $y(x)$ vanishes again at $x = x_2 > x_1$ with $p_1(x) > p(x)$ on (x_1, x_2), $z(x)$ will also vanish on the interval $x_1 < x < x_2$. For, consider a second solution $z_1(x)$ of (5.3), defined by the conditions $z_1(x_1) = 0$, $z_1'(x_1) = 1$. By the foregoing, $z_1(x)$ has a zero x_3 on the interval $x_1 < x < x_2$, and applying the Sturm separation theorem, we conclude that $z(x)$ has a zero on the interval $x_1 < x < x_3$.

Exercises

1.   Show that if $p(x) \leq 0$ on $[a, b]$, no nonnull solution of (5.1) can have more than one zero on $[a, b]$. (*Hint.* Compare with solutions of $[r(x)y']' = 0$.)

2.   If nonnull solutions $y(x)$ and $z(x)$, respectively, of the equations

$$x^2 y'' + xy' + (x^2 - 1)y = 0,$$

$$xz'' + z' + xz = 0$$

vanish at $x = 1$, which solution will vanish first after $x = 1$?

3.   Which of the following differential equations possesses the more rapidly oscillating solutions on the interval $(1, \infty)$:

$$x^2 y'' + xy' + y = 0,$$
$$y'' + y = 0?$$

(*Note.* Theorem 5.1 does not apply.)

4.   Which of the following differential equations possesses the more rapidly oscillating solutions on the interval $(1, \infty)$:

$$y'' + (x + 1)y = 0,$$
$$y'' + \sqrt{4 + x^2}\, y = 0?$$

5.   Are the solutions of the differential equation

$$(2x + 1)y'' + (x + 1)y = 0$$

oscillatory on the interval $(0, \infty)$?

6. The solutions of the differential equation

$$\left(\frac{x^2}{1 + x^2}\, y'\right)' + x \sin \frac{1}{x}\, y = 0$$

are oscillatory on the interval $(1, \infty)$. Estimate roughly the number of zeros a nonnull solution has on the interval $(10\pi, 20\pi)$.

7. If $q(x)$ and $f(x)$ are continuous and $q(x) > 0$ on the interval $a \le x \le b$, prove that the system

(1) $y'' - q(x)y = f(x)$,

                                  $y(a) = y(b) = 0$

has a unique solution. (*Hint.* Recall that the general solution of (1) can be written in the form $y_0(x) + c_1 u(x) + c_2 v(x)$, where $y_0(x)$ is any particular solution of (1) and $u(x)$ and $v(x)$ are linearly independent solutions of $y'' - q(x)y = 0$.)

Answers

3. The second equation.

5. Yes.

6. 10 or 11.

A generalization. We continue with the proof of the following theorem.

Theorem 5.2. If $p(x)$ and $q(x)$ are continuous on the interval $[a, b]$ with $p(x) \ne q(x)$, if $z(x)$ is a nonnull solution of the system

$$z'' + q(x)z = 0,$$
(5.5)
$$z(a) = z(b) = 0,$$

and if

(5.6) $$\int_a^b [p(x) - q(x)]z^2(x)\, dx \ge 0,$$

a nonnull solution $y(x)$ of the system

$$y'' + p(x)y = 0,$$
(5.7)
$$y(a) = 0$$

has a zero on the interval $a < x < b$.

Before proving the theorem we introduce two lemmas.

Lemma 5.1. If two differential equations

$$[r(x)y']' + a(x)y = 0,$$
$$[r(x)z']' + b(x)z = 0,$$

where $r(x) > 0$ and $r(x)$, $a(x)$, and $b(x)$ are continuous on I, have a common solution $u(x) \not\equiv 0$, then $a(x) \equiv b(x)$ on I.

To prove the lemma, observe that

$$[r(x)u'(x)]' + a(x)u(x) \equiv 0,$$
$$[r(x)u'(x)]' + b(x)u(x) \equiv 0.$$

It follows that $[a(x) - b(x)]u(x) \equiv 0$ and, hence, that $a(x) \equiv b(x)$ on I.

Lemma 5.2. If $y(x)$ and $z(x)$ are functions of class C' on I, with $y(x) \neq 0$ there, and if

$$y(x)z'(x) - y'(x)z(x) \equiv 0 \qquad (on\ I),$$

then $z(x) \equiv cy(x)$, where c is a constant.

By hypothesis, we may write

$$0 \equiv \frac{yz' - y'z}{y^2} \equiv \left(\frac{z}{y}\right)'.$$

Accordingly, $z/y \equiv c$, and the lemma is proved.

To prove the theorem consider the identities

$$z[z'' + qz] \equiv 0,$$
(5.8)
$$\frac{z^2}{y}[y'' + py] \equiv 0$$

on $[a, b]$. Note that the quotient y''/y is continuous on $[a, b]$, when defined suitably at a zero of y. Adding these identities and integrating, we have

(5.9)
$$\int_a^b \frac{z}{y}(yz' - zy')'\, dx = \int_a^b (p - q)z^2\, dx.$$

Now suppose that $y(x) \neq 0$ on $(a, b]$, and integrate the left-hand member of (5.9) by parts. Then,

$$(5.10) \qquad \left[\frac{z}{y}(yz' - zy') \right]_a^b - \int_a^b \left(z' - \frac{z}{y} y' \right)^2 dx = \int_a^b (p - q) z^2 \, dx.$$

The ratio z/y remains finite at $x = a$ (by l'Hospital's rule). The bracket in (5.10) is then zero. The integral

$$\int_a^b \left(z' - \frac{z}{y} y' \right)^2 dx$$

is positive; otherwise, $z(x) \equiv cy(x)$ (by Lemma 5.2), and $p(x) \equiv q(x)$ (by Lemma 5.1).

The assumption that $y(x) \neq 0$ on $(a, b]$ has led to a contradiction, and $y(x)$ must have a zero on $(a, b]$. Suppose now that $y(x) \neq 0$ on (a, b) and $y(b) = 0$. The ratio z/y in (5.10) would be finite at both $x = a$ and $x = b$, the bracket would again be zero, and again we would have a contradiction. Thus $y(x)$ must vanish on (a, b).

The proof of the theorem is complete.

Theorem 5.2 is readily seen to be a generalization of Theorem 5.1 when $r(x) \equiv 1$. An important special case of Theorem 5.2 occurs when $q(x) = \lambda^2$, where λ is a positive constant. In that case, condition (5.6) becomes

$$(5.11) \qquad \int_a^{a + \pi/\lambda} [p(x) - \lambda^2] \sin^2 \lambda(x - a) \, dx \geq 0.$$

This can be written

$$(5.12) \qquad \int_a^{a + \pi/\lambda} p(x) \sin^2 \lambda(x - a) \, dx \geq \frac{\pi}{2} \lambda.$$

If in (5.12) one sets $t = \lambda(x - a)$, one has the following alternate form of this condition:

$$(5.13) \qquad \int_0^\pi p \left(\frac{t}{\lambda} + a \right) \sin^2 t \, dt \geq \frac{\pi}{2} \lambda^2.$$

Example. Consider the differential system

$$(5.14) \qquad \begin{aligned} y'' + x^2 y &= 0, \\ y(0) &= 0. \end{aligned}$$

Here $p(x) = x^2$, $a = 0$, and condition (5.11) becomes

$$(5.15) \qquad \int_0^{\pi/\lambda} (x^2 - \lambda^2) \sin^2 \lambda x \, dx \geq 0.$$

Carrying out the indicated integration in (5.15) we have that

$$\lambda^4 \le \frac{2\pi^2 - 3}{6}.$$

That is, a (nonnull) solution $y(x)$ of (5.14) has a zero on the interval $0 < x < \dfrac{\pi}{\lambda_0}$, where

$$\lambda_0 = \left(\frac{2\pi^2 - 3}{6}\right)^{1/4}.$$

A little computation indicates that $y(x)$ has a zero on the interval

$$0 < x < 2.44.$$

Exercises

1. Using Theorem 5.2, find an upper bound for the first positive zero of a solution $y(x)$ of the differential system

    (a)
    $$y'' + (4 + x^2)y = 0,$$
    $$y(0) = 0;$$

    (b)
    $$y'' + (7 - x^2)y = 0,$$
    $$y(0) = 0;$$

    (c)
    $$y'' + xy = 0,$$
    $$y(0) = 0.$$

2. Find an upper bound for the smallest zero larger than 1 of a nonnull solution of the system

    $$y'' + \frac{1}{x}y = 0,$$
    $$y(1) = 0.$$

Answers

1. (a) 1.48;
    (b) Zero of solution is $\sqrt{6}/2$, exactly;
    (c) 2.7.

6   The Sturm-Picone theorem

The Sturm theorem may be generalized to apply to a more general pair of self-adjoint equations as follows.

Consider a pair of self-adjoint differential equations

(6.1) $[r(x)y']' + p(x)y = 0,$

(6.2) $[r_1(x)z']' + p_1(x)z = 0,$

where $r(x)$ and $r_1(x)$ are positive and $r(x)$, $p(x)$, $r_1(x)$, $p_1(x)$ are continuous on an interval $[a, b]$. The so-called *Picone formula*

(6.3) $\displaystyle\int_a^x [(r_1 - r)z'^2 + (p - p_1)z^2]\, dx$

$$+ \int_a^x r\left[\frac{yz' - y'z}{y}\right]^2 dx = \left[\frac{z}{y}(r_1 yz' - ry'z)\right]_a^x$$

$(a < x < b)$ may be derived along lines that yielded (5.10). A simpler proof (once the formula is known) is to differentiate both members of (6.3). The quantities y and z in (6.3) are solutions of (6.1) and (6.2), respectively, and the formula will be valid except possibly in the zeros of $y(x)$.

We shall consider nonnull solutions $y(x)$ and $z(x)$ such that

(6.4)
$$y(a) = 0,$$
$$z(a) = z(b) = 0.$$

Writing the Picone formula with the upper limit x replaced by b and supposing that $y(x) \neq 0$ on $(a, b]$, we have

$$\int_a^b [(r_1 - r)z'^2 + (p - p_1)z^2]\, dx + \int_a^b r\left[\frac{yz' - zy'}{y}\right]^2 dx = 0,$$

by an argument similar to that used in the proof of Theorem 5.2.

Theorem 6.1. Let $y(x)$ and $z(x)$ be solutions of (6.1) and (6.2), respectively, subject to conditions (6.4). If

$$\int_a^b [(r_1 - r)z'^2 + (p - p_1)z^2]\, dx > 0,$$

$y(x)$ must have a zero on the interval (a, b).

To prove the theorem suppose first that $y(x) \neq 0$ on $(a, b]$. The ratio z/y is well-defined on $[a, b]$, and (6.3) yields an immediate contradiction. Similarly, if $y(b)$ were zero, the ratio z/y would be continuous on $[a, b]$, and (6.3) yields a contradiction.

Corollary. If $r_1(x) \geq r(x)$ and $p(x) \geq p_1(x)$ with strict inequality holding in each of these conditions at at least one point of the interval, the solution $y(x)$ must vanish on (a, b).

The corollary is known as the *Sturm-Picone theorem.*

Exercise.   Prove formula (6.3).

7   The Bôcher-Osgood theorem

Consider the differential equation

(7.1) $$y'' + p(x)y = 0,$$

where $p(x)$ is positive and of class C' on the interval $I : 0 \leq x < \infty$. If $p'(x) \geq 0$ on I, then $p(x) \geq p(0) > 0$, and every solution of (7.1) vanishes infinitely often on I, by Sturm's comparison theorem. It follows from a result due to Bôcher and Osgood that under these conditions the amplitudes of the oscillation of solutions never increase as x increases on I.

Theorem 7.1. THE BÔCHER-OSGOOD THEOREM. *Suppose that $p(x)$ is of class C' and that $p(x) > 0, p'(x) \geq 0$ on $I : 0 \leq x < \infty$, and let $y(x)$ be an arbitrary solution of (7.1). If $x = a$ and $x = b$ are two consecutive zeros of $y'(x)$, then $|y(b)| \leq |y(a)|$.*

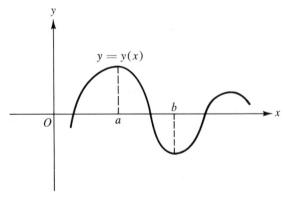

FIG. 5.4

Note that if $p'(x) \equiv 0$ for $a \leq x \leq b$, then $p(x)$ is constant on this interval and $|y(b)| = |y(a)|$. The theorem is accordingly true in this case. We may then proceed with the proof when $p'(x) \not\equiv 0$ on $[a, b]$.

Because $y(x)$ is a solution of (7.1), we have

(7.2) $y''(x) + p(x)y(x) \equiv 0.$

We multiply both members of (7.2) by $2y'(x)$ and integrate the result over the interval $[a, b]$ obtaining

(7.3) $y'^2(x) \Big|_a^b + \int_a^b p(x)[y^2(x)]'\, dx = 0.$

The first term of (7.3) is zero, and after an integration by parts we have

(7.4) $p(b)y^2(b) - p(a)y^2(a) = \int_a^b p'(x)y^2(x)\, dx.$

Suppose now that $y^2(b) > y^2(a)$. We shall show that this supposition leads to a contradiction. In that case,

$$\int_a^b p'(x)y^2(x)\, dx < y^2(b) \int_a^b p'(x)\, dx = y^2(b)[p(b) - p(a)],$$

and, consequently, from (7.4) we have

$$p(b)y^2(b) - p(a)y^2(a) < y^2(b)[p(b) - p(a)];$$

that is,

$$p(a)y^2(b) < p(a)y^2(a).$$

From this contradiction we infer the truth of the theorem.

The Bôcher-Osgood theorem could have been proved directly for a self-adjoint differential equation

(7.5) $[r(x)y']' + p(x)y = 0,$

where $r(x) > 0$, $r(x)$ and $p(x)$ are of class C' on I. The theorem in this case would be the following.

Theorem 7.2. If $r(x)$ and $p(x)$ are positive, and if $[r(x)p(x)]' \geq 0$, the conclusion of Theorem 7.1 is valid.

To prove the theorem we shall employ a transformation of (7.5) that is frequently useful. We set

$$t = \int_0^x \frac{dx}{r(x)}.$$

Then

$$r(x)y' = \frac{dy}{dt}, \qquad [r(x)y']' = \frac{d^2y}{dt^2}\frac{1}{r(x)},$$

and equation (7.5) becomes

(7.6) $$\ddot{y} + [p(x)r(x)]y = 0.$$

We see that the conditions of Theorem 7.1 will be met if

$$\frac{d}{dt}[p(x)r(x)] \geq 0 \qquad (0 \leq t < T_0),$$

where

$$T_0 = \int_0^\infty \frac{dx}{r(x)} \leq \infty.$$

But this derivative may be written as

$$[p(x)r(x)]'r(x),$$

and the theorem follows.

It follows from Theorem 7.2 that if $[r(x)p(x)]' \geq 0$, the solutions of (7.5) are bounded. When this inequality is reversed the solutions are not necessarily bounded. The following result is, however, noteworthy.

Theorem 7.3. If $r(x)$ and $p(x)$ are positive and of class C' on I and if

$$[r(x)p(x)]' \leq 0,$$

the products $r(x)p(x)y^2(x)$ and $r^2(x)y'^2(x)$, where $y(x)$ is any solution of (7.5), are bounded.

To prove the theorem we let y be a solution of (7.5) and multiply both members of (7.5) by $2ry'$. Integration yields the result that

$$(ry')^2\Big|_a^x + \int_a^x rp(y^2)'\,dx = 0.$$

An integration by parts permits us to write

$$[r(x)y'(x)]^2 + r(x)p(x)y^2(x) = c^2 + \int_a^x [r(x)p(x)]'y^2(x)\,dx,$$

where c^2 is a positive constant. Thus, each of $[r(x)y'(x)]^2$ and $r(x)p(x)y^2(x)$ is less than c^2, and the theorem is proved.

Exercises

1. Discuss Theorem 7.2 when

$$\int_0^\infty \frac{dx}{r(x)} < \infty.$$

2. Discuss the amplitudes of the solutions of the differential equation

$$\left(\frac{1}{x}y'\right)' + 4xy = 0 \quad (1 \le x < \infty).$$

3. Find a differential equation in the form (7.5) which has the solutions

$$\frac{1}{x^\alpha}\sin x, \ \frac{1}{x^\alpha}\cos x, \ \alpha \text{ constant},$$

and show that Theorem 7.2 applies when $\alpha \ge 1$. Note that the conclusion of Theorem 7.2 is valid for the function

$$\frac{1}{x^\alpha}\sin (x - \alpha), \quad \text{for } \alpha \ge 0.$$

4. Let $y^2(x)$ be the square of an arbitrary solution of the differential equation $y'' + y = 0$ and find a linear differential equation of third order of which $y^2(x)$ is a solution.

5. Find a linear differential equation of third order that is satisfied by the square of an arbitrary solution of equation (7.1). Assume that $p(x)$ is of class C'.

Answers

3. $(x^{2\alpha}y')' + [x^{2\alpha} + \alpha(\alpha - 1)x^{2\alpha - 2}]y = 0.$

5. $z''' + 4pz' + 2p'z = 0.$

8 A special pair of solutions

Consider the differential equation

(8.1) $[r(x)y']' + p(x)y = 0,$

where $r(x) > 0$, and $r(x)$ and $p(x)$ are continuous on some interval $I : a \le x < \infty$. In the sections that follow, we shall need to recall that if $y_1(x)$ is a solution of (8.1), a second linearly independent solution $y_2(x)$ is given by (cf. Chapter 4, Section 4)

$$(8.2) \qquad y_2(x) = y_1(x) \int_a^x \frac{dx}{r(x)\,y_1^2(x)}.$$

It will be remembered that the right-hand member of (8.2) may be defined in each zero $x = x_0$ of $y_1(x)$ as

$$\lim_{x \to x_0} \left[y_1(x) \int_a^x \frac{dx}{r(x)\,y_1^2(x)} \right].$$

The solution $y_2(x)$ will then be defined for all $x \geq a$.

9   Oscillation on a half-axis

In this section we shall develop a fundamental theorem that will provide a sufficient condition that solutions of an equation

$$(9.1) \qquad [r(x)y']' + p(x)y = 0$$

vanish infinitely often on the interval $0 < x < \infty$.

Theorem 9.1. Let $r(x)$ be positive, and suppose that $r(x)$ and $p(x)$ are continuous on the interval $0 < x < \infty$. If the two improper integrals

$$(9.2) \qquad \int_1^\infty \frac{dx}{r(x)} = +\infty, \qquad \int_1^\infty p(x)\,dx = +\infty,$$

then every solution $y(x)$ of (9.1) vanishes infinitely often on the interval $1 < x < \infty$. Similarly if the integrals

$$(9.3) \qquad \int_0^1 \frac{dx}{r(x)} = +\infty, \qquad \int_0^1 p(x)\,dx = +\infty,$$

every solution of (9.1) vanishes infinitely often on the interval $0 < x < 1$.

We shall prove the first statement of the theorem. † The second will be left to the student as an exercise.

Suppose that some solution $y(x)$ is nonoscillatory on $(1, \infty)$. Then, by the Sturm separation theorem, all (nonnull) solutions are nonoscillatory, and there exists a number $a > 1$ such that $y(x) \neq 0$ on $[a, \infty)$. The substitution

$$z = \frac{r(x)y'(x)}{y(x)} \qquad (a \leq x < \infty)$$

---

† A number of proofs of this theorem have been given. The present proof is due to W. J. Coles, "A simple proof of a well-known oscillation theorem," *Proc. Amer. Math. Soc.*, Vol. 19 (1968), p. 507.

leads to the identity

(9.4) $$z'(x) + \frac{z^2(x)}{r(x)} + p(x) \equiv 0 \qquad (a \le x < \infty)$$

that $z(x)$ satisfies. Integration of (9.4) from a to x $(>a)$ yields

(9.5) $$z(x) + \int_a^x \frac{z^2(x)}{r(x)} \, dx = z(a) - \int_a^x p(x) \, dx.$$

For x sufficiently large, say $x \ge b > a$, the right-hand member of (9.5) is negative. Accordingly, $z(x)$ is negative, for $x \ge b$, and

(9.6) $$z^2(x) > \left[\int_a^x \frac{z^2(x)}{r(x)} \, dx \right]^2.$$

Writing

$$I(x) = \int_a^x \frac{z^2(x)}{r(x)} \, dx,$$

we have from (9.6) that

$$r(x)I'(x) > I^2(x) \qquad (x \ge b).$$

Thus

$$\int_b^x \frac{I'(x)}{I^2(x)} \, dx > \int_b^x \frac{dx}{r(x)},$$

or,

$$\frac{1}{I(b)} > \frac{1}{I(x)} + \int_b^x \frac{dx}{r(x)} \qquad (x > b).$$

This contradicts (9.2), and the proof is complete.

Example. In the equation $y'' + a^2 y = 0$ $(a \ne 0)$, $r(x) = 1$, $p(x) = a^2$, conditions (9.2) are satisfied, and all solutions vanish infinitely often on the interval $1 < x < \infty$. This was already known to us, since a solution of the differential equation is $\sin ax$.

Example. In the differential equation

$$(xy')' + \frac{1}{x} y = 0,$$

we note that $r(x) = x$, $p(x) = \dfrac{1}{x}$. Both conditions (9.2) and (9.3) are satisfied. Accordingly, all solutions vanish infinitely often on $(1, \infty)$ and also on $(0, 1)$. It is easy to verify that $\sin \log x$ is a solution of this differential equation on the interval $(0, \infty)$.

If we were to consider the differential equation

$$\text{(9.7)} \qquad\qquad y'' + \frac{a^2}{x^2} y = 0,$$

we would note that

$$\int_1^\infty \frac{dx}{r(x)} = +\infty, \qquad \int_1^\infty p(x)\, dx < +\infty.$$

The test would fail. Since the differential equation is of Euler type we may solve it. It will be seen that solutions are *oscillatory* (that is, have an infinity of zeros) on $(1, \infty)$ when $a^2 > \frac{1}{4}$, and are *nonoscillatory* when $a^2 \leq \frac{1}{4}$.

The reason the test fails is that for this equation $[r(x) = 1]$ the integral $\int_1^x \dfrac{dx}{r(x)}$ becomes infinite too rapidly. To overcome this difficulty, we may transform (9.7) by means of the substitution

$$y = x^{1/2} z$$

obtaining

$$\text{(9.8)} \qquad\qquad (xz')' + \frac{a^2 - \frac{1}{4}}{x} z = 0.$$

Solutions z of (9.8) will be oscillatory if and only if solutions y of (9.7) are oscillatory. Theorem 9.1 applied to (9.8) yields the result that the solutions of (9.8) and hence those of (9.7) are oscillatory if $a^2 > \frac{1}{4}$. If $a^2 = \frac{1}{4}$, all solutions of (9.8) and of (9.7) are nonoscillatory. It follows then from the Sturm comparison theorem that all solutions of both equations also are nonoscillatory when $a^2 < \frac{1}{4}$.

It is frequently helpful, if the test of Theorem 9.1 fails to apply to an equation in the form (9.1), to try the substitution

$$\text{(9.9)} \qquad\qquad y = \left(\frac{x}{r}\right)^{1/2} z,$$

and to attempt to apply the test given by the theorem to the resulting differential equation in z.

For completeness we add the following result.

Theorem 9.2. Every nonnull solution of equation (9.1) *has at most a finite number of zeros on the interval* $a \le x < \infty$, *if*

$$\int_a^\infty \frac{dx}{r(x)} < +\infty \quad \text{and} \quad \left| \int_a^x p(x)\, dx \right| < M \qquad (a \le x < \infty),$$

where M is any positive constant.

The proof is omitted.†

Exercises

1. Solve the differential equation $x^2 y'' + a^2 y = 0$. (Separate cases according as $a^2 > \frac{1}{4}$ or $\le \frac{1}{4}$.)

2. Show that solutions of the differential equation

    $$x^p y'' + k^2 y = 0 \quad (p \text{ constant}; k^2 > 0)$$

    are oscillatory on the interval $(1, \infty)$ if and only if either $p < 2$ or $p = 2$, $k^2 > \frac{1}{4}$. (*Hint.* Consider separately the following cases: $p \le 1$, $1 < p < 2$, $p = 2$, $p > 2$.)

3. Show that the substitution $y = u(x)z$ in the differential equation

    $$[r(x)y']' + p(x)y = 0$$

    yields the equation

    $$(ru^2 z')' + u[(ru')' + pu]z = 0.$$

4. Use Theorem 9.1 to test Bessel's equation,

    $$x^2 y'' + xy' + (x^2 - n^2)y = 0 \qquad (n \text{ constant}).$$

5. It is known that for each integer $n \ge 0$ Laguerre's equation

    $$xy'' + (1 - x)y' + ny = 0$$

    has a polynomial solution of degree n. It follows that no solution of this differential equation can be oscillatory on the interval $(1, \infty)$. Find these polynomial solutions when $n = 0, 1, 2, 3$. Then use the theory of this chapter to show that if n is any constant, the solutions of Laguerre's equation are nonoscillatory on the interval $(1, \infty)$. [*Hint.* Use (9.9).]

---

† R. A. Moore, "The Behavior of Solutions of a Linear Differential Equation of Second Order." *Pac. Jour. Math.*, **5**, p. 135 (1955).

6. Do the same as in Exercise 5 for Hermite's equation

$$y'' - 2xy' + 2ny = 0.$$

[*Hint.* Substitute $y = e^{x^2/2}z$.]

7. Show that all solutions of the equation $(x^q y')' + x^q y = 0$ (*q constant*) are oscillatory on the interval $(1, \infty)$.

8. By solving the differential equation

$$x^2 y'' + axy' + by = 0 \qquad (a, b \text{ constants}),$$

one observes that its solutions are oscillatory on the interval $(1, \infty)$ and on the interval $(0, 1)$, if and only if $(a - 1)^2 - 4b < 0$. Use the methods of this chapter to prove this result for the interval $(1, \infty)$.

9. Test the equation

$$xy'' + \left(1 + \frac{1}{\log x}\right)y' + y = 0$$

for oscillation of solutions on the interval $(2, \infty)$.

10. Prove that if

$$\int_1^\infty \left[xp(x) - \frac{1}{4x}\right] dx = +\infty,$$

the solutions of $y'' + p(x)y = 0$ are oscillatory on $[1, \infty)$. Show that the solutions are nonoscillatory on $(1, \infty)$ if $4x^2 p(x) - 1 \le 0$, for x large. Assume $p(x)$ continuous on $[1, \infty)$.

11. Prove that if

$$\int_0^1 \left[xp(x) - \frac{1}{4x}\right] dx = +\infty,$$

the solutions of $y'' + p(x)y = 0$ are oscillatory on $(0, 1]$. Show that if $4x^2 p(x) - 1 \le 0$, for x positive and sufficiently small, the solutions are nonoscillatory on $(0, 1)$. Assume $p(x)$ continuous on $(0, 1]$.

12. Prove the final statement in Theorem 9.1.

10 Two transformations

It is frequently useful to transform an equation of the type

(10.1) $$[r(x)y']' + p(x)y = 0$$

into an equation of the form

(10.2) $$y'' + P(x)y = 0.$$

We assume that $r(x)$, $r'(x)$, $r''(x)$, and $p(x)$ are continuous, and $r(x) > 0$ in $[a, b]$. There are two methods commonly used. In the first, we set

(10.3) $$y = u(x)z \qquad [u(x) > 0]$$

and obtain the equation

(10.4) $$(ru^2z')' + u[(ru')' + pu]z = 0$$

(see Exercise 3 of Section 9). If we let $ru^2 = 1$, or

$$u = [r(x)]^{-1/2},$$

we shall have

$$z'' + P(x)z = 0,$$

where

$$P(x) = u[(ru')' + pu].$$

Equation (10.3) was a transformation of the dependent variable. An equation of the type (10.2) can also be obtained by transforming the independent variable. In this case, set

(10.5) $$t = \int_a^x \frac{dt}{r(t)},$$

and note that

$$\frac{dt}{dx} = \frac{1}{r(x)}.$$

Because $r(x) > 0$, t is a strictly increasing function of x, and equation (10.5) also defines x as an increasing function of t. Call this function $g(t)$. Then,

$$y' = \frac{dy}{dx} = \frac{dy}{dt} \cdot \frac{dt}{dx} = \frac{1}{r(x)} \frac{dy}{dt} = \frac{1}{r(x)} \dot{y},$$

and

$$(ry')' = \frac{d}{dx} \dot{y} = \frac{d}{dt} (\dot{y}) \frac{dt}{dx} = \ddot{y} \frac{1}{r(x)}.$$

Equation (10.1) then becomes

(10.6) $$\ddot{y} + rpy = 0,$$

where the product rp in (10.6) means

$$r[g(t)]p[g(t)].$$

Example.   Consider the differential equation

$$x^2y'' - 2xy' + (2 + x^2)y = 0 \qquad (x > 0).$$

If we put this equation in self-adjoint form, we have

(10.7) $$\left(\frac{1}{x^2}y'\right)' + \frac{2 + x^2}{x^4}y = 0.$$

Here,

$$r = \frac{1}{x^2}, \qquad p = \frac{2 + x^2}{x^4},$$

and the transformation (10.3) becomes

$$y = xz.$$

An easy computation yields

$$u[(ru')' + pu] = 1,$$

and equation (10.4) becomes

$$z'' + z = 0.$$

Linearly independent solutions of (10.7) are, accordingly,

$$x \sin x, \qquad x \cos x.$$

If, however, we apply the transformation (10.5) to (10.7), we have†

$$t = \int_0^x x^2 \, dx = \frac{x^3}{3},$$

and equation (10.6) becomes

―――――――――

† A cautious reader may wish to replace the lower limit on the integral sign by a positive constant.

$$\ddot{y} + \frac{2 + x^2}{x^6} y = 0,$$

or

(10.8) $$\ddot{y} + \frac{2 + (3t)^{2/3}}{9t^2} y = 0.$$

In this example, the transformation (10.3) of the dependent variable led to an easy solution of (10.7), while (10.8) appears to be less interesting. In practice, however, both transformations are useful, and either may have advantages over the other in a given situation.

Exercises

1. Transform the equation

$$x^2 y'' - 2mxy' + [m(m + 1) + x^2]y = 0$$

by means of (10.3), and thus find two linearly independent solutions of the equation.

2. Employ the transformation (10.5) to solve the differential equation

$$[r(x)y']' + \frac{1}{r(x)} y = 0,$$

where $r(x) > 0$ and continuous on the interval $[a, b]$.

3. Solve the differential equation

$$u'(x)z'' - u''(x)z' + u'^3(x)z = 0,$$

where $u'(x) > 0$ and $u''(x)$ is continuous on an interval $a < x < b$.

4. Use (10.4) to solve the differential equation

$$[w^2(x)y']' + w(x)[w''(x) + w(x)]y = 0,$$

where $w(x) \neq 0$ and of class C'' on (a, b).

5. If $u'(x) > 0$, $w(x) \neq 0$, and $u(x)$ and $w(x)$ are of class C'' on (a, b), show that linearly independent solutions of the differential equation

$$\left[\frac{w^2(x)}{u'(x)} y'\right]' + \left[\left[\frac{w'(x)}{u'(x)}\right]' + u'(x)w(x)\right]w(x)y = 0$$

are

$$\frac{\sin u(x)}{w(x)} \quad \text{and} \quad \frac{\cos u(x)}{w(x)}.$$

6.   Show that linearly independent solutions of the differential equation

$$x^2 y'' + (2m - n + 1)xy' + [n^2 x^{2n} + m(m - n)]y = 0 \qquad (x > 0)$$

are

$$\frac{\sin x^n}{x^m} \quad \text{and} \quad \frac{\cos x^n}{x^m},$$

provided $n \neq 0$.

7.   Use the result in Exercise 6 to solve the differential equation

$$x^2 y'' - 4xy' + (x^2 + 6)y = 0 \qquad (x > 0).$$

8.   Use the result in Exercise 6 to solve the differential equation

$$x^2 y'' + xy' + (4x^4 - 1)y = 0 \qquad (x > 0).$$

9.   By a suitable choice of the function $u(x)$, transform the differential equation

$$y'' + b(x)y' + c(x)y = 0$$

by means of the substitution

$$y = u(x)z$$

into a differential equation of the form

$$z'' + P(x)z = 0.$$

Assume $b'(x)$ and $c(x)$ continuous on an interval $a \leq x \leq b$.

10.   Given the differential equation

(1) $$y'' + b(x)y' + c(x)y = 0,$$

where $b(x)$ and $c(x)$ are continuous on $I : [a, b]$, show that every solution that vanishes at a point $x = x_0$ of I is a constant multiple of the solution

(2) $$y_2(x_0)y_1(x) - y_1(x_0)y_2(x),$$

where $y_1(x)$ and $y_2(x)$ are any linearly independent solutions of (1). Then, considering x_0 as a variable parameter, show that

$$\frac{\partial}{\partial x_0}[y_2(x_0)y_1(x) - y_1(x_0)y_2(x)] = y_2'(x_0)y_1(x) - y_1'(x_0)y_2(x)$$

is a solution of (1) linearly independent of (2).

Answers

1. $x^m \sin x$, $x^m \cos x$.

3. $c_1 \sin u(x) + c_2 \cos u(x)$.

4. $\quad c_1 \dfrac{\sin x}{w(x)} + c_2 \dfrac{\cos x}{w(x)}.$

11 More on oscillation

Consider the differential equation

(11.1) $[r(x)y']' + p(x)y = 0,$

where $r(x) > 0$, and $r(x)$, $p(x)$ are continuous on the interval $I:[a, \infty)$. If $v(x)$ and $w(x)$ are any two solutions of (11.1), it is not difficult to verify that

(11.2) $u(x) = \sqrt{v^2(x) + w^2(x)}$

is a solution of the differential equation

(11.3) $ru^3[(ru')' + pu] = k^2 \qquad (a \le x < \infty),$

where

$$r[v(x)w'(x) - w(x)v'(x)] \equiv k.$$

Thus, when $v(x)$ and $w(x)$ are linearly independent, $k \neq 0$, and $u(x) \neq 0$ on I, inasmuch as $v(x)$ and $w(x)$ cannot then have a common zero.

Note, conversely, that the differential equation (11.3) with $k = 1$

(11.4) $ru^3[(ru')' + pu] = 1$

always has a solution $u(x) \neq 0$ on I [replace $v(x)$ by $v(x)/k$, for example].

We recall that if $u(x) \neq 0$ is any function of class C' on I, and if we make the substitution $y = u(x)z$ in (11.1), we obtain [see (10.4)] the differential equation

(11.5) $(ru^2z')' + u[(ru')' + pu]z = 0$

for z. Further, if $u(x)$ is such that

(11.6) $(ru^2)u[(ru')' + pu] = 1,$

linearly independent solutions of (11.5) are

$$\sin \int_a^x \frac{dx}{r(x)u^2(x)}, \qquad \cos \int_a^x \frac{dx}{r(x)u^2(x)}.$$

But (11.6) is simply (11.4). Inasmuch as solutions of (11.1) are oscillatory on

I, if and only if solutions of (11.5) are oscillatory there, it follows that if solutions of (11.1) are oscillatory on I, there exists a function $u(x) \neq 0$ of class C' on I such that

$$(11.7) \qquad \int_a^\infty \frac{dx}{ru^2} = +\infty, \qquad \int_a^\infty u[(ru')' + pu]\, dx = +\infty.$$

On the other hand, if there exists a function $u(x) \neq 0$ of class C' on I such that conditions (11.7) hold, it follows from Theorem 9.1 that solutions of (11.1) are oscillatory on I. We have proved the following result.

Theorem 11.1. A necessary and sufficient condition that solutions of (11.1) be oscillatory is that there exist a function $u(x) \neq 0$ of class C' on I for which conditions (11.7) hold.

Theorem 11.1 provides a matrix for a large number of sufficient conditions both for oscillation and nonoscillation on I. For example, suppose $r(x) = 1$ and $a = 1$. Equation (11.1) becomes

$$(11.8) \qquad y'' + p(x)y = 0,$$

and conditions (11.7) become

$$(11.9) \qquad \int_1^\infty \frac{dx}{u^2} = +\infty, \qquad \int_1^\infty u(u'' + pu)\, dx = +\infty.$$

If we set $u = x^\alpha$, these conditions become

$$(11.9)' \quad \int_1^\infty \frac{dx}{x^{2\alpha}} = +\infty, \qquad \int_1^\infty [x^{2\alpha}p(x) + \alpha(\alpha - 1)x^{2\alpha - 2}]\, dx = +\infty.$$

From the first conditions in (11.9)$'$ we note that $2\alpha \leq 1$.

Accordingly, *if there exists a constant $\alpha \leq \frac{1}{2}$ such that*

$$(11.10) \qquad \int_1^\infty x^{2\alpha}p(x) + \alpha(\alpha - 1)x^{2\alpha - 2}]\, dx = +\infty,$$

the solutions of (11.8) are oscillatory.

In particular, if $\alpha = \frac{1}{2}$, we obtain the criterion in Exercise 10 of Section 9. If $r(x) = 1$, $p(x) = \sin x$, equation (11.8) becomes

$$(11.11) \qquad y'' + (\sin x)y = 0.$$

We try $u = 2 + \sin x$ in (11.9) and observe that the first condition is satisfied. The second condition becomes

$$\int_1^\infty (2 + \sin x)[-\sin x + \sin x(2 + \sin x)]\, dx = +\infty.$$

This condition will be satisfied if

$$\lim_{x \to \infty} \int_1^x \sin x(1 + \sin x)\, dx = +\infty,$$

which is readily verified. Solutions of (11.11) are, accordingly, oscillatory.

Finally, the differential equation

$$(11.12) \qquad\qquad y'' + \frac{k}{x^3} y = 0 \qquad (1 \le x < \infty),$$

for example, may be managed by determining a number α such that

$$u\left(u'' + \frac{k}{x^3} u\right) < 0 \qquad \text{(for } x \text{ large)}$$

where $u = x^\alpha$. Such a choice is $\alpha = \frac{1}{2}$. Thus, solutions of (11.12) are non-oscillatory on $[1, \infty)$.

Theorem 9.2 may also be applied to equation (11.12) along with the theory above.

The following result is an immediate consequence of (11.5).

Theorem 11.2. If there exists a positive function $u(x)$ of class C'' on $[a, \infty)$ such that $(ru')' + pu < 0$, for x large, the solutions of (11.1) are nonoscillating on that interval.

Exercises

1. Verify equation (11.3).

2. Are the solutions of $y'' + (1 + \sin x)y = 0$ oscillatory on $(0, \infty)$?

3. Discuss the oscillation of solutions of $y'' + (\cos 2x)y = 0$.

4. Show that the solutions of the differential equation

$$y'' + \left[\frac{1}{4x^2} + \frac{k}{x^2 \log^2 x}\right] y = 0 \qquad (e \le x < \infty)$$

are oscillatory for $k > \frac{1}{4}$ and nonoscillatory for $k \le \frac{1}{4}$. [*Hint.* An educated guess as to a function $u(x)$ will solve the problem; otherwise, solve the differential equation. The substitution $x = e^t$ followed by $y = e^{t/2} z$ will be helpful.]

5.   Show that the solutions of

$$(x^\beta y')' + k^2 x^\beta y = 0$$

are oscillatory on $[1, \infty)$ for every constant β, and each constant $k \neq 0$.

6.   Let $r(x) > 0$ and continuous on $[1, \infty)$. Discuss the oscillation of solutions of

$$\left[\frac{1}{r(x)}\, y'\right]' + r(x)y = 0$$

on $[1, \infty)$.

12   Liapunov's inequality

This famous inequality is given in (12.2) below.

Theorem 12.1. If $p(x)$ is continuous on the interval I:$[a, b]$ and if there exists a solution $y(x) \not\equiv 0$ of the differential equation

(12.1) $$y'' + p(x)y = 0$$

that has two zeros on I, then

(12.2) $$\int_a^b p_+(x)\, dx > \frac{4}{b-a}.$$

Here, $p_+(x) = \frac{1}{2}[p(x) + |p(x)|]$.

Consider the differential equation

(12.3) $$y'' + p_+(x)y = 0.$$

Inasmuch as $p_+(x) \geq p(x)$ on $[a, b]$, there exists a solution $y(x)$ of (12.3) such that $y(a) = y(c) = 0$, with $y(x) > 0$ on $[a, c] \subset I$. Let $x = f$ be a point on (a, c) at which $y(x)$ attains its absolute maximum. We note that since $y''(x) \leq 0$ on (a, c), $y(x)$ possesses no relative minima on that (open) interval (it may possess a line of relative maxima, however). Draw the two chords that connect the point $P[f, y(f)]$ with the points $(a, 0)$ and $(c, 0)$.
Then, a reference to Fig. 5.5 makes it plain that

$$y'(a) > \frac{y(f)}{f-a} \qquad \text{and} \qquad -y'(c) > \frac{y(f)}{c-f} \geq \frac{y(f)}{b-f}.$$

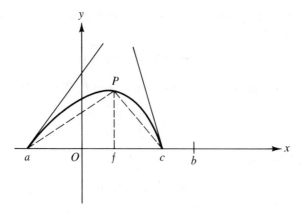

FIG. 5.5

It follows that

(12.4)
$$-\frac{y'(c) - y'(a)}{y(f)} > \frac{b - a}{(f - a)(b - f)};$$

that is,

$$-\frac{1}{y(f)} \int_a^c y''(x)\, dx = \frac{1}{y(f)} \int_a^c p_+(x) y(x)\, dx > \frac{b - a}{(f - a)(b - f)}.$$

Inasmuch as

$$\frac{1}{y(f)} \int_a^c p_+(x) y(x)\, dx < \int_a^c p_+(x)\, dx,$$

we have

(12.5)
$$\int_a^b p_+(x)\, dx \geq \int_a^c p_+(x)\, dx > \frac{b - a}{(f - a)(b - f)}.$$

But the last member of (12.5) is readily seen to be greater than or equal to $4/(b - a)$, with equality only if f is the midpoint of (a, b).

The proof is complete.†

---

† This method of proof is substantially that of Hartman and Wintner, "On an oscillation criterion of Liapounoff," *Amer. Jour. Math.*, Vol. 73 (1951), p. 888.

This inequality may be extended readily to the general self-adjoint differential equation

(12.6) $$[r(x)y']' + p(x)y = 0,$$

where $r(x) > 0$ and $r(x)$, $p(x)$ are continuous on I. We employ the substitution

(12.7) $$t = \int_a^x \frac{dx}{r(x)},$$

and (12.6) becomes

(12.8) $$\ddot{y} + r(x)p(x)y = 0.$$

If a solution $y(x)$ of (12.6) has consecutive zeros $x = a$ and $x = c$, there is a solution $y_1(t)$ of (12.8) with consecutive zeros at $t = 0$ and at $t = t_0$, where

$$t_0 = \int_a^c \frac{dx}{r(x)}.$$

According to Theorem 12.1 applied to (12.8), we then have

$$\int_0^{t_0} r(x)p_+(x)\, dt > \frac{4}{t_0}.$$

Evaluation of this integral by means of the substitution (12.7) yields

(12.9) $$\int_a^c \frac{dx}{r(x)} \int_a^c p_+(x)\, dx > 4.$$

The number 4 in (12.2) and in (12.9) may be shown to be a "best" constant; that is, it cannot be replaced by a larger number.

13   Gronwall's inequality

In this section we shall develop a generalization and an extension of what is known as Gronwall's inequality. This inequality is frequently used to establish the uniqueness of a solution of a system of differential equations.

Suppose that $u(t)$ and $v(t)$ are continuous and $c(t)$ is of class C' on an interval $I:[a, b]$.

Theorem 13.1.† *If $v(t) \geq 0$ and $c'(t) \geq 0$ on I, and if*

---

† Theorem 13.1 appears to have been first proved by Reid [see W. T. Reid, *Ordinary Differential Equations*, Wiley-Interscience, New York (1971)].

(13.1) $u(t) \leq c(t) + \displaystyle\int_a^t v(t)u(t)\, dt \qquad (a \leq t \leq b),$

then

(13.2) $u(t) \leq c(t) \exp \displaystyle\int_a^t v(t)\, dt.$

To prove this, set

$$q(t) = \int_a^t v(t)u(t)\, dt.$$

Then

(13.3) $q'(t) = v(t)u(t) \leq v(t)c(t) + v(t)q(t),$

and

$$q'(t) - v(t)q(t) \leq c(t)v(t).$$

Multiplying both members of this inequality by $\exp\left[-\int_a^t v(t)\, dt\right]$ (and changing t to s) we have

(13.4) $\dfrac{d}{ds}\left\{ q(s) \exp\left[-\displaystyle\int_a^s v(t)\, dt \right] \right\} \leq c(s)v(s) \exp\left[-\displaystyle\int_a^s v(t)\, dt \right].$

If we integrate both members of (13.4) from a to t and recall that $q(a) = 0$, we have

$$q(s) \exp\left[-\int_a^s v(t)\, dt \right]\Bigg|_{s=a}^{s=t} \leq \int_a^t c(s)v(s) \exp\left[-\int_a^s v(t)\, dt \right] ds,$$

or

(13.5) $q(t) \leq \displaystyle\int_a^t c(s)v(s) \exp\left[\int_s^t v(t)\, dt \right] ds.$

After an integration by parts and applying (13.1), we have

(13.6) $u(t) \leq c(a) \exp\left[\displaystyle\int_a^t v(t)\, dt \right] + \displaystyle\int_a^t c'(s) \exp\left[\int_s^t v(t)\, dt \right] ds.$

Since $c'(s) \geq 0$ on I, it follows that

$$u(t) \leq c(a) \exp\left[\int_a^t v(t)\, dt\right] + \int_a^t c'(s) \exp\left[\int_a^t v(t)\, dt\right] ds$$

(13.7)
$$\leq c(t) \exp \int_a^t v(t)\, dt.$$

When $c(t)$ is a constant, (13.7) becomes Gronwall's inequality.
The proof of Theorem 13.1 is complete.

Theorem 13.2. If $v(t) \geq 0$, $c'(t) \leq 0$ on I, and if

$$u(t) \geq c(t) + \int_a^t v(t)u(t)\, dt,$$

then

(13.8)
$$u(t) \geq c(t) \exp \int_a^t v(t)\, dt.$$

Corollary. If $c(t)$ is a constant c,

$$u(t) \geq c \exp \int_a^t v(t)\, dt.$$

The proof is strictly analogous to the proof of Theorem 13.1. In place of (13.3) we have

$$q'(t) \geq v(t)c(t) + v(t)q(t).$$

The inequality (13.6) becomes

$$u(t) \geq c(a) \exp\left[\int_a^t v(t)\, dt\right] + \int_a^t c'(s) \exp\left[\int_s^t v(t)\, dt\right] ds,$$

and because $v(t) \geq 0$ and $c'(t) \leq 0$ on I, we have

$$u(t) \geq c(a) \exp\left[\int_a^t v(t)\, dt\right] + \int_a^t c'(s) \exp\left[\int_a^t v(t)\, dt\right] ds,$$

and (13.8) is then immediate.†

---

† In Theorems 13.1 and 13.2, the given conditions on $c(t)$ may, it will be seen, be replaced by the conditions that $c(t)$ be an increasing, or decreasing, function, respectively. Such a function will have a derivative almost everywhere, and if one regards the integrals as Lebesgue integrals, the analysis in the text will remain valid.

Example.   Suppose it is known that a function $u(t)$ satisfies the inequality

(13.9) $$u(t) \le t + \int_0^t u(t)\, dt \qquad (t \ge 0).$$

Theorem (13.1) then yields the fact that

(13.10) $$u(t) \le te^t.$$

The solution $e^t - 1$ of the differential equation

$$u' - u = 1$$

clearly satisfies (13.9). The inequality (13.10) then yields the fact that

$$e^t - 1 \le te^t \qquad (t \ge 0).$$

6

Characteristic functions; orthogonal polynomials

1 Introduction

In this chapter we shall develop an introduction to the theory of *characteristic functions* (also known as *eigenfunctions*) and their associated *eigenvalues* (also known as *characteristic numbers* and as *eigennumbers*).

Consider the differential system

(1.1)
$$y'' + \lambda y = 0,$$
$$y(0) = 0, \qquad y(\pi) = 0,$$

where λ is a parameter independent of x. We seek solutions $y(x) \not\equiv 0$ of this system. It is clear at the outset that such a solution will exist only for certain values of λ. Indeed, the general solution of the differential equation is

$$c_1 \sin \sqrt{\lambda}\, x + c_2 \cos \sqrt{\lambda}\, x.$$

Since $y(0) = 0$, $c_2 = 0$. Thus, if there is a solution $y(x) \not\equiv 0$ of (1.1), it must be of the form

(1.2)
$$y(x) = c_1 \sin \sqrt{\lambda}\, x \qquad (c_1 \neq 0).$$

Applying the condition $y(\pi) = 0$ to (1.2), we note that $\sqrt{\lambda}\,\pi$ must be an integral multiple of π. That is,

$$\lambda = n^2,$$

where n is an integer $\neq 0$. In other words, λ must be one of the numbers

128

(1.3)   1, 4, 9, ...

Corresponding solutions $y(x)$ are, respectively,

(1.4) $\sin x, \ \sin 2x, \ \sin 3x, \ \ldots$.

The numbers (1.3) are the *characteristic numbers*, and the set of functions
(1.4) is a corresponding set of *characteristic functions*. Clearly, we might
replace the set (1.4) by any set

$$a_1 \sin x, \ a_2 \sin 2x, \ a_3 \sin 3x, \ \ldots,$$

where each of the constants $a_1, a_2, a_3, \ldots$ may be given any value different
from zero.

It will be observed that for this problem there is a smallest characteristic
number ($\lambda = 1$), that there is an infinite sequence of characteristic numbers,
and that the limit of this sequence is $+\infty$.

More generally, we consider the differential system

(1.5)
$$[r(x)y']' + p(x)y + \lambda q(x)y = 0,$$

$$y(a) = 0, \qquad y(b) = 0,$$

where $r(x)$ and $q(x)$ are positive and $r(x)$, $q(x)$, and $p(x)$ are continuous real
functions of x on the closed interval $a \leq x \leq b$. The constant λ is again a
parameter independent of x. We seek solutions $y(x) \not\equiv 0$ of (1.5). The
system (1.5) is an example of a *Sturm-Liouville* system. A value of λ for
which (1.5) has a solution $y(x, \lambda)$ is a characteristic number. The corresponding function $y(x, \lambda)$ is the characteristic function.

To simplify the analysis, we assume that no nonnull solution of the system

(1.5)′
$$[r(x)y']' + p(x)y = 0,$$

$$y(a) = 0$$

vanishes on $(a, b]$. It follows from the Sturm comparison theorem that if λ
is a number for which a nonnull solution of (1.5) exists, λ must be positive.

*Theorem 1.1. Subject to the above assumption on the system (1.5), there
exists an infinite sequence of characteristic numbers $\lambda_1, \lambda_2, \ldots$ of (1.5) with the
properties*

$$0 < \lambda_1 < \lambda_2 < \cdots, \qquad \lambda_n \to +\infty,$$

and a corresponding sequence of characteristic functions

$$y_1(x), \ y_2(x), \ \ldots$$

*defined on the interval $a \leq x \leq b$. The function $y_n(x)$ has precisely n zeros on
the interval $a < x \leq b$.*

To prove the theorem let us first, for convenience, extend the definitions of $r(x)$, $p(x)$, $q(x)$ to the interval $[a, \infty)$ by defining

$$r(x) = r(b), \qquad p(x) = p(b), \qquad q(x) = q(b)$$

on (b, ∞). These functions will then retain the properties specified in the theorem on the larger interval.

Because the system $(1.5)'$ has only the null solution, for some (small) positive value of λ, $\lambda = \lambda_0$, there will exist a nonnull solution $y(x, \lambda_0)$ of the differential equation with the property that

$$y(a, \lambda_0) = 0, \qquad y(x_0, \lambda_0) = 0$$

where $x_0 > b$ is the first zero following $x = a$ of the solution $y(x, \lambda_0)$. We shall observe the behavior of this zero x_0 as λ is increased steadily always requiring that $y(a, \lambda) = 0$.

Note that by the Sturm comparison theorem [because $q(x) > 0$], as λ increases, x_0 decreases, and also by the Sturm comparison theorem, a small increase in λ causes a small decrease in x_0. That is, x_0 is a continuous, strictly decreasing function of λ. Accordingly, there exists a value λ_1 $(0 < \lambda_1 < \lambda_0)$ and a solution $y(x, \lambda_1)$ such that

$$y(a, \lambda_1) = 0, \qquad y(b, \lambda_1) = 0,$$

with $y(x, \lambda_1) \neq 0$ on (a, b). The only additional observation to ensure the validity of this statement is that for λ sufficiently large, the first zero following $x = a$ of a solution $y(x, \lambda)$ that vanishes at $x = a$ can be taken arbitrarily close to $x = a$—again, by the Sturm comparison theorem and the fact that $q(x) > 0$.

The number λ_1 is the first *eigenvalue* of the system (1.5), and the solution $y_1(x, \lambda_1) = y(x, \lambda_1)$ is the corresponding (*first*) *characteristic function* (also *eigenfunction*).

Next, we allow λ to increase steadily from the value λ_1. The zero of $y_1(x, \lambda)$ that was at $x = b$ moves into the interval (a, b), and a second zero $x = x_1$ will appear on the interval $(b, b + \varepsilon)$ $(\varepsilon > 0$, small). Repeating the earlier argument, we see that there will be determined a number λ_2 $(0 < \lambda_1 < \lambda_2)$ such that there exists a corresponding solution $y_2(x, \lambda_2)$ with the property that

$$y_2(a, \lambda_2) = y_2(c, \lambda_2) = y_2(b, \lambda_2) = 0 \qquad (a < c < b),$$

with $y_2(x, \lambda_2) \neq 0$ on (a, b), except at $x = c$. The number λ_2 is the second eigenvalue of the system (1.5), and $y_2(x, \lambda_2)$ is the corresponding characteristic function.

The process can be continued, and we shall have constructed an infinite sequence of eigenvalues

$$0 < \lambda_1 < \lambda_2 < \cdots$$

and a corresponding sequence of characteristic functions

$$y_1(x, \lambda_1), y_2(x, \lambda_2), \ldots$$

with the property that $y_n(x, \lambda_n)$ vanishes at $x = a$, $x = b$, and at $n - 1$ points $c_1 < c_2 < \cdots < c_{n-1}$ on (a, b) and is not equal to zero elsewhere on (a, b). It follows from the existence theorem that all the zeros of $y_n(x, \lambda_n)$ are simple zeros. It is clear that the sequence $\lambda_1, \lambda_2, \ldots$ cannot be bounded, for with λ sufficiently large, say $\lambda > \lambda^*$, the quantity $p(x) + \lambda q(x) > 1$ for all x on $[a, \infty)$, and all solutions will be oscillatory on $[a, \infty)$. Consider any value of λ larger simultaneously than λ^* and a supposed upper bound of the sequence $\{\lambda_n\}$. A solution $y(x, \lambda)$ vanishing at $x = a$ will have a zero following $x = b$. By increasing λ, that zero would move into the point $x = b$, and the contradiction would follow (see Fig. 6.1).

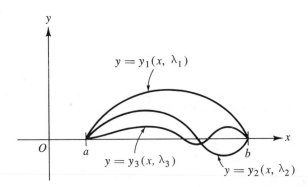

FIG. 6.1

The proof of the theorem is complete.

Critique. It should be noted that in the proof of the theorem the notation is somewhat oversimplified. As λ changes, the solution, subject to the condition that it vanishes at $x = a$, also changes with λ. A more accurate, but more cumbersome notation for such a solution would be $y_\lambda(x, \lambda)$. Indeed, absolute precision would also require that $y_\lambda(x, \lambda)$ always be subject to an additional condition such as

$$\frac{\partial y_\lambda(a, \lambda)}{\partial x} = 1.$$

This last condition is omitted because all nonnull solutions that vanish at $x = a$ have the same zeros on (a, ∞).

Finally, the proof of continuity of x_0 as a function of λ can be given analytically as follows. Let $y(a, x, \lambda)$ be a nonnull solution that vanishes at $x = a$ and at $x = a_0$ when $\lambda = \lambda_*$. This solution is of class C' in x on $[a, \infty)$ and analytic in λ for $-\infty < \lambda < \infty$. We need to solve the equation

$$y(a, x, \lambda) = 0$$

for x. This is possible, by the implicit-function theorem, inasmuch as $y(a, x_0, \lambda_*) = 0$, $y_x(a, x_0, \lambda_*) \neq 0$ (because the solution is $\not\equiv 0$), and the solution function $x(\lambda)$ is at least C' in λ.

Returning to the example (1.1), note that the system corresponding to (1.5)' becomes

$$y'' = 0,$$
$$y(0) = 0, \qquad y(\pi) = 0,$$

and that the only solution of this system is the null solution. Accordingly, the theorem applies.

Exercises

Determine the sequences of characteristic numbers and functions for the following Sturm-Liouville systems.

1. $y'' + \lambda y = 0,$
    $y(0) = 0,$
    $y(\pi/2) = 0.$

2. $y'' + \lambda y = 0,$
    $y(-\pi/2) = 0,$
    $y(\pi/2) = 0.$

3. $y'' + y + \lambda y = 0,$
    $y(0) = 0,$
    $y(\pi) = 0.$

4. $y'' + \lambda y = 0,$
    $y(0) = 0,$
    $y(1) = 0.$

5. $(xy')' + \lambda \dfrac{1}{x} y = 0,$
    $y(1) = y(e^\pi) = 0.$

6. $(x^2y')' + \lambda \dfrac{1}{x^2} y = 0,$

     $y(1) = y(2) = 0.$

7. $x^2y'' + \lambda y = 0,$

     $y(1) = y(e^\pi) = 0.$

8. $y'' - 2y' + \lambda y = 0,$

     $y(0) = y(1) = 0.$

9. $x^2y'' + xy' + (x^2 - \tfrac{1}{4})y = 0,$

     $y(\pi/2) = y(3\pi/2) = 0.$ [*Hint.* See Ex. 6, Sec. 10, Ch. 11.]

Answers

1. $\lambda = 4n^2,\, y_n(x) = \sin 2nx \qquad (n = 1, 2, 3, \dots).$

2. $\lambda = n^2,\, y_n(x) = \sin n(x + \pi/2) \qquad (n = 1, 2, 3, \dots).$

3. $\lambda = n^2 - 1,\, y_n(x) = \sin nx \qquad (n = 1, 2, 3, \dots).$

4. $\lambda = n^2\pi^2,\, y_n(x) = \sin n\pi x \qquad (n = 1, 2, 3, \dots).$

5. $\lambda = n^2,\, y_n(x) = \sin (n \log x) \qquad (n = 1, 2, 3, \dots).$

6. $\lambda = 4n^2\pi^2,\, y_n(x) = \sin 2n\pi[1 - (1/x)] \qquad (n = 1, 2, 3, \dots).$

7. $\lambda = n^2 + \tfrac{1}{4},\, y_n(x) = x^{1/2} \sin (n \log x).$

2 Orthogonality

Two functions $f(x)$ and $g(x)$ are said to be *orthogonal* on the interval (a, b) with respect to the *weight* function $q(x)$ if

$$\int_a^b q(x)f(x)g(x)\, dx = 0.$$

Theorem 2.1. Each characteristic function associated with the Sturm-Liouville system (1.5) is orthogonal to every other characteristic function on the interval (a, b), with respect to the weight function q(x).

Let $y_n(x)$ and $y_m(x)$ be arbitrary characteristic functions, and let λ_n and λ_m be, respectively, their associated eigenvalues. We have to show that

(2.1) $$\int_a^b q(x)y_n(x)y_m(x)\, dx = 0 \qquad (n \neq m).$$

To prove (2.1), we observe that

$$[r(x)y'_n(x)]' + [p(x) + \lambda_n q(x)]y_n(x) \equiv 0,$$
(2.2)
$$[r(x)y'_m(x)]' + [p(x) + \lambda_m q(x)]y_m(x) \equiv 0,$$

since $y_n(x)$ and $y_m(x)$ are solutions of the system (1.5) when $\lambda = \lambda_n$ and $\lambda = \lambda_m$, respectively. Let us multiply both members of the first identity by $y_m(x)$, those of the second by $-y_n(x)$, and add. We obtain

(2.3) $y_m(x)[r(x)y'_n(x)]' - y_n(x)[r(x)y'_m(x)]' \equiv (\lambda_m - \lambda_n)q(x)y_n(x)y_m(x).$

If both members of the identity (2.3) are integrated from a to b, and if integration by parts is applied to the left-hand member, there results

(2.4) $r(x)[y_m(x)y'_n(x) - y_n(x)y'_m(x)]_a^b = (\lambda_m - \lambda_n)\int_a^b q(x)y_n(x)y_m(x)\,dx.$

The left-hand member of (2.4) is clearly zero, and since $\lambda_m \neq \lambda_n$ when $m \neq n$, equation (2.1) follows at once.

The sequence of characteristic functions is said to be an *orthogonal sequence*. If, in addition,

$$\int_a^b q(x)y_n^2(x)\,dx = 1 \qquad (n = 1, 2, \ldots),$$

the sequence $y_1(x), y_2(x), \ldots$ is said to be *normal*. Recall that if $y_n(x)$ is a characteristic function, so is $a_n y_n(x)$, where a_n is any constant different from zero. Accordingly, we can normalize the sequence of characteristic functions by means of the following device. Suppose that

$$\int_a^b q(x)y_n^2(x)\,dx = c_n^2 \qquad (n = 1, 2, 3, \ldots).$$

Since $c_n^2 > 0$, the sequence

$$\frac{y_1(x)}{c_1}, \frac{y_2(x)}{c_2}, \ldots$$

is immediately seen to be both normal and orthogonal with respect to the weight function $q(x)$ on the interval (a, b).

In the case of the example (1.1), the sequence $\{\lambda_n\}$ was seen to be given by $\lambda_n = n^2$, and a corresponding sequence of characteristic functions is

$$\sin x, \sin 2x, \sin 3x, \ldots.$$

Inasmuch as $q(x) = 1$, and

$$\int_0^\pi \sin^2 nx\,dx = \frac{\pi}{2},$$

a normal, orthogonal sequence of characteristic functions on $(0, \pi)$ is

$$\frac{\sin x}{c}, \frac{\sin 2x}{c}, \frac{\sin 3x}{c}, \ldots,$$

where $c = \sqrt{\pi/2}$.

Exercises

Using the first three computed characteristic functions for the various exercises at the end of Section 1, verify directly that these three functions are orthogonal in pairs. Find also the associated normal characteristic functions.

3   The expansion of a function in a series of orthogonal functions

Thousands of pages have been written on this subject and its ramifications. We shall have to be content here with an indication as to how the formal expansion of a given function may be achieved.

Let $y_1(x)$, $y_2(x), \ldots$ be a sequence of normal characteristic functions of (1.5), and suppose that $f(x)$ is a given function continuous on the interval $a \le x \le b$ which may be represented there by a series $\sum_1^\infty b_n y_n(x)$, where the b_n are constants. Proceeding formally, we have†

$$f(x) = b_1 y_1(x) + b_2 y_2(x) + \cdots = \sum_{n=1}^\infty b_n y_n(x),$$

$$q(x)f(x)y_m(x) = b_1 q(x)y_1(x)y_m(x) + b_2 q(x)y_2(x)y_m(x) + \cdots$$

$$= \sum_{n=1}^\infty b_n q(x)y_n(x)y_m(x),$$

(3.1) $$\int_a^b q(x)f(x)y_m(x)\, dx = \int_a^b b_1 q(x)y_1(x)y_m(x)\, dx$$

$$+ \int_a^b b_2 q(x)y_2(x)y_m(x)\, dx + \cdots$$

$$= \int_a^b \left[\sum_{n=1}^\infty b_n q(x)y_n(x)y_m(x) \right] dx$$

---

† The manipulations which follow would be valid if we assumed, for example, that the series converged uniformly to $f(x)$ on the interval.

$$= \sum_{n=1}^{\infty} b_n \int_a^b q(x)\,y_n(x)\,y_m(x)\,dx.$$

From (3.1) we have

(3.2) $$\int_a^b q(x)f(x)y_m(x)\,dx = b_m \qquad (m = 1, 2, \ldots).$$

Thus, if there is a series expansion of the type assumed, then

$$f(x) = b_1 y_1(x) + b_2 y_2(x) + \cdots,$$

where the numbers b_m are determined by (3.2). Having determined the form of the coefficients b_m, we may now present the basic problem of expanding a given function in a series of normal characteristic functions. Let $f(x)$ be the given function, which we suppose for simplicity to be continuous on the interval $a \le x \le b$, and define numbers b_m by the equations

$$b_m = \int_a^b q(x)f(x)y_m(x)\,dx \qquad (m = 1, 2, \ldots).$$

Then the formal expansion of $f(x)$ is given by

(3.3) $$f(x) \sim b_1 y_1(x) + b_2 y_2(x) + \cdots.$$

The symbol " $\sim$ " is used rather than " $=$ " since we have no *a priori* assurance that the series will converge or, if it is convergent, that it will converge to the function $f(x)$. Indeed, it may not converge, and if it is convergent, it need not converge to $f(x)$. We may then ask what conditions (in addition to continuity) must be imposed upon $f(x)$ to ensure that the series actually converges to $f(x)$ on $[a, b]$; when will the convergence be uniform; when will it be absolute; and so on. New concepts of convergence arise in which the theory of the Lebesgue integral plays a fundamental role. For completeness we state without proof the following theorem.†

Theorem 3.1. If $f(x)$ is continuous and if $f'(x)$ exists and is piecewise continuous‡ on the interval $a \le x \le b$, the series (3.3) converges uniformly and absolutely to $f(x)$ on every closed interval interior to the open interval $a < x < b$.

---

† See E. L. Ince, *Ordinary Differential Equations*, p. 273, Longmans, Roberts and Green, London (1927).

‡ A function is *piecewise continuous* on an interval $[a, b]$ if the interval can be divided into a finite number of subintervals on the interior of each of which the function is continuous and if it possesses finite limits as the ends of the subintervals are approached from the interiors of the subintervals.

Exercises

1. Expand the following functions defined on $(0, \pi)$ in a series of characteristic functions of the system

$$y'' + \lambda y = 0,$$
$$y(0) = y(\pi) = 0.$$

(a) x;   (b) x^2;   (c) e^x.

2. Do the same as in Exercise 1, when the interval is $(-\pi/2, \pi/2)$ and the boundary conditions are $y(-\pi/2) = y(\pi/2) = 0$.

3. Let the system be

$$(xy')' + \lambda \frac{1}{x} y = 0,$$
$$y(1) = y(e^\pi) = 0,$$

and expand formally the function x^{-2} in a series of characteristic functions on the interval $(1, e^\pi)$.

4. The Legendre differential equation

$$[(1 - x^2)y']' + n(n + 1)y = 0 \qquad (n = 0, 1, 2, \ldots)$$

is known to have polynomial solutions $P_n(x)$ of degree n. Compute $P_n(x)$ when $n = 0, 1, 2, 3, 4$, subject to the condition that $P_n(1) = 1$. Show that the sequence $P_1(x), P_2(x), \ldots$ is an orthogonal sequence of functions on $(-1, 1)$ with weight function unity.

5. Obtain the first few terms in the expansion of e^x in a series of Legendre polynomials on the interval $(-1, 1)$.

Answers

1. (a) $2(\sin x - \frac{1}{2}\sin 2x + \frac{1}{3}\sin 3x - \cdots)$;

(b) $2\left[\left(\pi - \frac{4}{\pi}\right)\sin x - \frac{\pi}{2}\sin 2x + \left(\frac{\pi}{3} - \frac{4}{3^3\pi}\right)\sin 3x - \frac{\pi}{4}\sin 4x + \cdots\right]$;

(c) $\sum_{1}^{\infty} a_n \sin nx, \ a_n = \frac{2n}{1 + n^2}\frac{1}{\pi}[1 - (-1)^n e^\pi]$.

3. $\sum_{1}^{n} a_n \sin(n \log x), \ a_n = \frac{2}{\pi}\frac{n}{n^2 + 4}[(-1)^{n+1}e^{-2\pi} + 1]$.

4. $P_0(x) = 1, \ P_1(x) = x, \ P_2(x) = \frac{1}{2}(3x^2 - 1), \ P_3(x) = \frac{1}{2}(5x^3 - 3x),$
    $P_4(x) = \frac{1}{8}(35x^4 - 30x^2 + 3).$

5. $\frac{1}{2}(e - e^{-1}) + 3e^{-1}x + \frac{5}{4}(e - 7e^{-1})(3x^2 - 1) + \cdots.$

4   The Bessel equation

The differential equation

$$(4.1) \qquad x^2 y'' + x y' + (x^2 - n^2)y = 0 \qquad (n \geq 0)$$

is known as *Bessel's equation*. It is of considerable importance in mathematical physics. It is also of interest in its own right to mathematicians. We shall examine some of the simpler properties of solutions of this equation.

First, we note that Bessel's equation has a singular point at $x = 0$—the only singular point on the interval $(-\infty, \infty)$. It can be shown that this equation always has a solution

$$(4.2) \qquad J_n(x) = x^n W_n(x),$$

where $W_n(x)$ is analytic on $(-\infty, \infty)$, and $W_n(0) \neq 0$. A second linearly independent solution may be obtained by the usual substitution

$$y = J_n(x)v.$$

Power series expansions of both solutions may be found in many places. Large treatises have been devoted only to *Bessel functions*, that is, to solutions of Bessel's equation.† We shall necessarily limit ourselves to a few elementary properties of solutions of this equation.

Bessel's equation may be put in self-adjoint form. We have

$$(4.3) \qquad (xy')' + \left(x - \frac{n^2}{x}\right)y = 0.$$

We observe that all solutions are oscillatory on the intervals $[\varepsilon, \infty)$ for all positive values of ε. [Solutions are also oscillatory on the intervals $(-\infty, -\varepsilon)$.]

If the substitution

$$y = x^{-1/2}z$$

is made in equation (4.1), this equation becomes

$$(4.4) \qquad z'' + \left(1 - \frac{n^2 - \frac{1}{4}}{x^2}\right)z = 0.$$

---

† The classic is G. N. Watson, *A Treatise on the Theory of Bessel Functions*, 2d ed., Cambridge, New York (1966); 804 pages.

When x is large, this equation is approximately

(4.5) $z'' + z = 0.$

Applying the Sturm comparison theorem, we see that when $n^2 < \frac{1}{4}$, the solutions of (4.4) oscillate more rapidly than do solutions of (4.5). When $n^2 > \frac{1}{4}$, the opposite is true. And when $n^2 = \frac{1}{4}$, equation (4.4) becomes equation (4.5), and the general solution of Bessel's equation in this case is

$$c_1 \frac{\sin x}{\sqrt{x}} + c_2 \frac{\cos x}{\sqrt{x}}.$$

The form of (4.4) suggests that when $x \to \infty$, *the distance between consecutive zeros of Bessel functions tends to π* (all n). This can be seen as follows. On the interval $[c, c + \pi]$, c large, the coefficient of z lies between

$$h^2 = 1 - \frac{n^2 - \frac{1}{4}}{c^2} \quad \text{and} \quad k^2 = 1 - \frac{n^2 - \frac{1}{4}}{(c + \pi)^2} \qquad (h > 0, \, k > 0).$$

Accordingly, a solution of (4.4) vanishing at $x = c$ will have its next zero between

(4.6) $c + \dfrac{\pi}{h} \quad \text{and} \quad c + \dfrac{\pi}{k} \qquad (c \text{ large}),$

by the Sturm comparison theorem. The solution $\sin(x - c)$ of

$$y'' + y = 0$$

will have its first zero following $x = c$ at $x = c + \pi$. The differences between $c + \pi$ and the terms in (4.6) are

$$\pi\left(1 - \frac{1}{h}\right) \quad \text{and} \quad \pi\left(1 - \frac{1}{k}\right).$$

Both these differences can be made arbitrarily small by taking c sufficiently large, and the italicized statement above has been established.

Exercise.   Consider a Bessel function that vanishes at $x = x_1 > 0$. Let $\delta(x_1)$ be the distance from x_1 to the next larger zero of this solution. Show that as x_1 increases, $\delta(x_1)$ is a strictly increasing function of x_1, when $n^2 < \frac{1}{4}$, and a strictly decreasing function, when $n^2 > \frac{1}{4}$. [The function $\delta(x_1)$ is called the *conjugacy function*.]

It can also be shown that $\delta(x_1)$ is a convex function, when $n^2 > \frac{1}{4}$ (and concave when $n^2 < \frac{1}{4}$), but the author is not aware of an elementary proof of this result.

Graphs of the functions $J_0(x)$ and $J_1(x)$ are given in Fig. 6.2. It should be remarked that $J_0(x)$ and $J_1(x)$ are each special solutions of their respective equations. They are known as *principal solutions* at $x = 0$ because it can be shown that

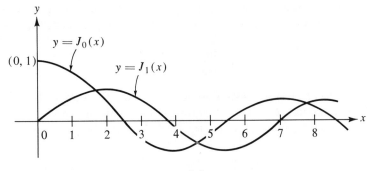

FIG. 6.2

$$(4.7) \qquad \lim_{x \to 0} \frac{J_0(x)}{B_0(x)} = 0, \qquad \lim_{x \to 0} \frac{J_1(x)}{B_1(x)} = 0,$$

when $B_0(x)$ and $B_1(x)$ are solutions of their respective equations that are linearly independent of their respective numerators in (4.7).

The function $W_n(x)$.   The substitution [recall (4.2)]

$$(4.8) \qquad y = x^n W \qquad (n \geq 0)$$

in the Bessel equation (4.1) yields the differential equation

$$(4.9) \qquad xW'' + (2n + 1)W' + xW = 0,$$

the self-adjoint form of which is

$$(4.10) \qquad (x^{2n+1}W')' + x^{2n+1}W = 0.$$

The differential equations (4.9) and (4.10) have an attractive symmetry. In effect, the transformation (4.8) has removed some of the singularity at $x = 0$ from Bessel's equation, and all the properties of Bessel functions can be derived from (4.9) or (4.10).

Note that replacing x by $-x$ in (4.9), and consequently, y' by $-y'$ and y'' by $-y''$, leaves the equation unaltered. Accordingly, all solutions $W(x)$ are *even functions*; that is, $W(-x) \equiv W(x)$.

Let us then attempt to find a power series solution $W_n(x)$ of (4.9) subject to the condition $W_n(0) = 1$. Inasmuch as such a solution would necessarily be even, we set

$$W_n(x) = 1 + \sum_{m=1}^{\infty} a_{2m} x^{2m}.$$

Then

$$W_n'(x) = \sum_{m=1}^{\infty} 2m a_{2m} x^{2m-1},$$

$$W_n''(x) = \sum_{m=1}^{\infty} 2m(2m-1) a_{2m} x^{2m-2}.$$

Substituting in (4.9), we have

$$(4.11) \quad \sum_{m=1}^{\infty} 2m(2m-1) a_{2m} x^{2m-1}$$

$$+ \sum_{m=1}^{\infty} (2n+1) 2m a_{2m} x^{2m-1} + x + \sum_{m=1}^{\infty} a_{2m} x^{2m+1} \equiv 0;$$

that is, we wish to determine the coefficient a_{2m} so that the left-hand member of (4.11) is a power series with zero coefficients. This will be true if

$$(4.12) \quad [2a_2 + 2(2n+1)a_2 + 1]x$$

$$+ \sum_{m=2}^{\infty} [(2m(2m-1) + 2m(2n+1))a_{2m} + a_{2m-2}] x^{2m-1} \equiv 0.$$

We have then the equation

$$(4.13) \quad a_2 = -\frac{1}{4(n+1)}, \qquad a_{2m} = -\frac{a_{2m-2}}{4m(m+n)} \qquad (m = 2, 3, 4, \ldots)$$

for the determination of the numbers a_{2m}. These numbers may be calculated consecutively from (4.13).

The test-ratio test applied to the resulting series yields

$$\left| \frac{a_{2m}}{a_{2m-2}} x^2 \right| = \left| \frac{1}{4m(m+n)} x^2 \right|,$$

and this ratio tends to zero as $m \to \infty$ for every x on $(-\infty, \infty)$. The series then converges for all x, and its sum represents a solution of the differential equation (4.9) on the entire x-axis. The series may be written in the form

$$(4.14) \qquad W_n(x) = 1 - \frac{x^2}{2^2 1!\,(n+1)} + \frac{x^4}{2^4 2!\,(n+1)(n+2)}$$

$$- \frac{x^6}{2^6 3!\,(n+1)(n+2)(n+3)} + \cdots,$$

or

$$(4.14)' \quad W_n(x) = 1 + \sum_{m=1}^{\infty} (-1)^m \frac{x^{2m}}{2^{2m} m!\,(n+1)(n+2)\cdots(n+m)}.$$

Such a series may be differentiated term by term, as often as desired, and it follows that

$$W_n'(x)$$

$$= \sum_{m=1}^{\infty} \frac{(-1)^m 2m x^{2m-1}}{2^{2m} m!\,(n+1)(n+2)\cdots(n+m)}$$

$$= \sum_{m=1}^{\infty} (-1)^m \frac{x^{2m-1}}{2^{2m-1}(m-1)!\,(n+1)(n+2)\cdots(n+m)}$$

$$= -\frac{x}{2(n+1)} + \sum_{m=2}^{\infty} (-1)^m \frac{x^{2m-1}}{2^{2m-1}(m-1)!\,(n+1)(n+2)\cdots(n+m)}$$

$$= -\frac{x}{2(n+1)} \left[1 + \sum_{m=2}^{\infty} (-1)^{m-1} \frac{x^{2m-2}}{2^{2m-2}(m-1)!\,(n+2)(n+3)\cdots(n+m)} \right]$$

$$= -\frac{x}{2(n+1)} \left[1 + \sum_{m=1}^{\infty} (-1)^m \frac{x^{2m}}{2^{2m} m!\,(n+2)(n+3)\cdots(n+m+1)} \right].$$

Thus, finally, we have

$$(4.15) \qquad W_n'(x) = -\frac{x}{2(n+1)} W_{n+1}(x).$$

We are prepared to prove the following theorem.

Theorem 4.1. If $x = a$ and $x = b$ are two consecutive positive zeros of a Bessel function $B_n(x)$ $(n > 0)$, each Bessel function $B_m(x)$ $(0 < m < n)$ has a zero on (a, b). A similar statement is true if a and b are negative.

This theorem is an immediate consequence of the Sturm comparison and separation theorems applied to equation (4.3).

If $m > n \geq 0$, it is clear that the zeros of $B_m(x)$ will not separate those of

$B_n(x)$, of course, but a limited statement of that type is true. We have the following result.

Theorem 4.2. The zeros of $W_{n+1}(x)$ separate those of $W_n(x)$ $(n \geq 0)$.

As in Theorem 4.1, n need not be an integer, of course.

To prove the theorem, let $x = a$ and $x = b$ be a pair of consecutive zeros of $W_n(x)$ with a and b having the same sign. By Rolle's theorem, $W'_n(x)$ has a zero on (a, b). But $W_{n+1}(x)$ has this same zero, by (4.15).

The Bessel functions designated by $J_n(x)$ $(n \geq 0)$ are traditionally defined by the relation

$$J_n(x) = c_n x^n W_n(x).$$

Accordingly, Theorems 4.1 and 4.2 apply to the functions $J_n(x)$, and we have the following corollary.

Corollary. The zeros of $J_n(x)$ separate those of $J_{n+1}(x)$ $(n \geq 0)$, and also those of $J_{n-1}(x)$ $(n \geq 1)$.

By "separation" is meant separating pairs of consecutive positive zeros and pairs of consecutive negative zeros of the functions $J_i(x)$.

Exercises

1. From equations (4.9) and (4.15), deduce the recursion formula.

$$x^2 W_{n+1} = 4n(n + 1)(W_n - W_{n-1}) \qquad (n \geq 1).$$

2. For $n \geq 1$ an integer, set $J_n = \frac{1}{2^n n!} x^n W_n$ in Exercise 1 and derive the recursion formula

$$J_{n+1} = \frac{2n}{x} J_n - J_{n-1}.$$

(This formula is also valid for all $n \geq 1$, whether or not n is an integer.)

3. Show that the general solution of $y'' + 9x^2 y = 0$ is given by

$$y = \sqrt{x}[c_1 J_{1/4}(\tfrac{3}{2}x^2) + c_2 J_{-1/4}(\tfrac{3}{2}x^2)].$$

4. Find the general solution of the differential equation $y'' + a^2 xy = 0$ $(x > 0)$.

5.   Show that

$$z = t^{1/2}J_n(kt^\alpha)$$

is a solution of

$$\ddot{z} + [\alpha^2 k^2 t^{2\alpha-2} + (\tfrac{1}{4} - \alpha^2 n^2)t^{-2}]z = 0.$$

5   Orthogonal polynomials

Earlier in this chapter we met several systems of orthogonal functions. In Exercise 4 of Section 3 we obtained the Legendre polynomials of degree 0, 1, 2, 3, 4 and observed that the Legendre polynomials form a sequence of functions orthogonal in the interval $(-1, 1)$ with respect to the weight function 1.

There are four other classical sequences of orthogonal polynomials given below, but, as we shall see, we may assume any positive continuous weight function $p(x)$ on an arbitrary interval (a, b) and construct a corresponding sequence of orthogonal polynomials.

CLASSICAL ORTHOGONAL POLYNOMIALS

SYSTEM	DIFFERENTIAL EQUATION	WEIGHT FUNCTION	INTERVAL
Legendre	$[(1 - x^2)y']' + n(n + 1)y = 0$	1	$(-1, 1)$
Chebyshev	$[(1 - x^2)^{1/2}y']'$ $+ n^2(1 - x^2)^{-1/2}y = 0$	$(1 - x^2)^{-1/2}$	$(-1, 1)$
Laguerre	$(xe^{-x}y')' + ne^{-x}y = 0$	e^{-x}	$(0, \infty)$
Hermite	$(e^{-x^2}y')' + 2ne^{-x^2}y = 0$	e^{-x^2}	$(-\infty, \infty)$
Jacobi	$[(1 - x)^{1+\alpha}(1 + x)^{1+\beta}y']'$ $+ n(n + \alpha + \beta + 1)$ $\times (1 - x)^\alpha(1 + x)^\beta y = 0$	$(1 - x)^\alpha$ $\times (1 + x)^\beta$	$(-1, 1)$ $(\alpha > -1,$ $\beta > -1)$

We have chosen to write the above differential equations in self-adjoint form, for the weight function can then be seen immediately. In addition, this form is the natural one to use in proving the polynomials orthogonal. (These equations will be found in expanded form in Exercise 1, Section 4, Chapter 5.)

For example, suppose $y_n(x)$ and $y_m(x)$ are polynomial solutions of Hermite's equation, where $n \neq m$. Since the coefficient of y, except for a constant factor, is e^{-x^2}, we anticipate that e^{-x^2} is the weight function. We have

$$[e^{-x^2}y_n'(x)]' + 2ne^{-x^2}y_n(x) \equiv 0,$$

$$[e^{-x^2}y_m'(x)]' + 2me^{-x^2}y_m(x) \equiv 0 \qquad (n \neq m).$$

Multiply both members of the first equation by $y_m(x)$ and both members of the second equation by $-y_n(x)$ and add. Then

$$(5.1) \quad y_m(x)[e^{-x^2}y_n'(x)]' - y_n(x)[e^{-x^2}y_m'(x)]' \equiv 2e^{-x^2}(m - n)y_n(x)y_m(x).$$

Next, integrate both members of (5.1) on the interval $(-\infty, \infty)$. We have, after an integration by parts,

$$e^{-x^2}[y_m(x)y_n'(x) - y_n(x)y_m'(x)]_{-\infty}^{\infty} = 2(m - n)\int_{-\infty}^{\infty} e^{-x^2}y_n(x)y_m(x)\,dx,$$

or

$$(5.2) \qquad 0 = 2(m - n)\int_{-\infty}^{\infty} e^{-x^2}y_n(x)y_m(x)\,dx.$$

Since $m - n \neq 0$, we see that the polynomials $y_n(x)$ and $y_m(x)$ are orthogonal on the interval $(-\infty, \infty)$ with respect to the weight function e^{-x^2}.

The Schmidt orthogonalization process. Suppose $p(x)$ is positive and continuous on the interval $[a, b]$. We shall construct a sequence of polynomials $P_0(x), P_1(x), P_2(x), \ldots$ orthogonal on this interval with respect to the weight function $p(x)$. The degree of $P_n(x)$ will be n.

We set

$$P_0(x) = 1,$$

and write $P_1(x) = a_1 x + b_1$, where the constants a_1 and b_1 are to be determined. Since $P_0(x)$ and $P_1(x)$ are to be orthogonal we have

$$\int_a^b p(x)(1)(a_1 x + b_1)\,dx = 0,$$

or

$$(5.3) \qquad a_1 k_1 + b_1 k_0 = 0,$$

where

$$k_1 = \int_a^b xp(x)\,dx,$$

$$k_0 = \int_a^b p(x)\,dx.$$

We may then write

$$P_1(x) = a_1\left(x - \frac{k_1}{k_0}\right),$$

and note that a_1 may be chosen as any number not equal to zero. For simplicity, we write

$$P_1(x) = \alpha_1 x + \beta_1 \qquad (\alpha_1 \neq 0).$$

Next, we write

$$P_2(x) = a_2 x^2 + b_2 x + c_2.$$

We wish to determine the constants $a_2 \neq 0, b_2, c_2$, such that P_2 is orthogonal to P_0 and to P_1. Then,

$$\int_a^b p(x)(1)(a_2 x^2 + b_2 x + c_2)\, dx = 0,$$

$$\int_a^b p(x)(\alpha_1 x + \beta_1)(a_2 x^2 + b_2 x + c_2)\, dx = 0.$$

If we set

$$k_2 = \int_a^b x^2 p(x)\, dx,$$

$$k_3 = \int_a^b x^3 p(x)\, dx,$$

we have the equations

(5.4)
$$a_2 k_2 + b_2 k_1 + c_2 k_0 = 0,$$
$$a_2(\alpha_1 k_3 + \beta_1 k_2) + b_2(\alpha_1 k_2 + \beta_1 k_1) + c_2(\alpha_1 k_1 + \beta_1 k_0) = 0$$

for the determination of a_2, b_2, c_2.

Solutions of (5.4) may be verified to be

(5.5)
$$a_2 = \alpha_1(k_1^2 - k_0 k_2),$$
$$b_2 = \alpha_1(k_0 k_3 - k_1 k_2),$$
$$c_2 = \alpha_1(k_2^2 - k_1 k_3).$$

All other solutions of these equations are proportional to the set (5.5). Recall that $\alpha_1 \neq 0$. Because we wish to be sure that the degree of $P_2(x)$ is actually 2, it remains to be shown that $k_1^2 - k_0 k_2 \neq 0$; that is, that the determinant

(5.6)
$$\begin{vmatrix} \int_a^b p(x)\, dx & \int_a^b xp(x)\, dx \\[2ex] \int_a^b xp(x)\, dx & \int_a^b x^2 p(x)\, dx \end{vmatrix} \neq 0.$$

Suppose the determinant were zero. There would then exist constants c_1 and c_2, not both zero, such that

$$c_1 \int_a^b p(x)\, dx + c_2 \int_a^b xp(x)\, dx = 0,$$

$$c_1 \int_a^b xp(x)\, dx + c_2 \int_a^b x^2 p(x)\, dx = 0.$$

We multiply the first equation by c_1, the second by c_2, and add. The result may be written in the form

$$\int_a^b [c_1 \sqrt{p(x)} + c_2 x \sqrt{p(x)}]^2\, dx = 0.$$

It would follow that

$$c_1 \sqrt{p(x)} + c_2 x \sqrt{p(x)} \equiv 0,$$

and hence, that

$$c_1 + c_2 x \equiv 0,$$

which contradicts the hypothesis that not both c_1 and c_2 are zero.

The process may be continued, and each polynomial will be determined in turn. In each case, we may argue as before that the leading coefficient of the polynomial is different from zero.

Exercises

1. Develop $P_0(x)$, $P_1(x)$, $P_2(x)$, $P_3(x)$ from the:

    (a) Chebyshev differential equation;
    (b) Laguerre differential equation;
    (c) Hermite differential equation.

    (*Suggestion.* Use the expanded forms of the differential equations and try for a power series solution.)

2. Prove that the Chebyshev polynomials form an orthogonal sequence.

3. Prove that the Laguerre polynomials form an orthogonal sequence.

4. Prove that the Jacobi polynomials form an orthogonal sequence.

5. Let $p(x)$ be positive and continuous on the interval $[a, b]$. Prove that the determinant

$$\begin{vmatrix} \int_a^b p(x)\,dx & \int_a^b xp(x)\,dx & \int_a^b x^2 p(x)\,dx \\ \int_a^b xp(x)\,dx & \int_a^b x^2 p(x)\,dx & \int_a^b x^3 p(x)\,dx \\ \int_a^b x^2 p(x)\,dx & \int_a^b x^3 p(x)\,dx & \int_a^b x^4 p(x)\,dx \end{vmatrix} \neq 0.$$

6. Given the weight function 1 and the interval $(-1, 1)$, use the Schmidt process of orthogonalization to derive $P_0(x)$, $P_1(x)$, $P_2(x)$, $P_3(x)$. (Your answers should be Legendre polynomials except for multiplicative constants.)

7. Discuss solutions of the Chebyshev differential equation in the light of Exercise 7 of Chapter 5, Section 4.

Answers

1. (a) $T_0(x) = 1$, $T_1(x) = x$, $T_2(x) = 2x^2 - 1$, $T_3(x) = 4x^3 - 3x$;

    (b) $L_0(x) = 1$, $L_1(x) = -x + 1$, $L_2(x) = x^2 - 4x + 2$,
        $L_3(x) = -x^3 + 9x^2 - 18x + 6$;

    (c) $H_0(x) = 1$, $H_1(x) = 2x$, $H_2(x) = 4x^2 - 2$, $H_3(x) = 8x^3 - 12x$.

7

Systems of linear differential equations

1 The existence theorem

In this chapter we study systems of linear differential equations; that is, systems of the type

$$\frac{dx_1}{dt} = a_{11}(t)x_1 + a_{12}(t)x_2 + \cdots + a_{1n}(t)x_n,$$

$$\frac{dx_2}{dt} = a_{21}(t)x_1 + a_{22}(t)x_2 + \cdots + a_{2n}(t)x_n,$$

(1.1)

$$\cdot \quad \cdot \quad \cdot \quad \cdot \quad \cdot \quad \cdot \quad \cdot \quad \cdot \quad \cdot \quad \cdot \quad \cdot \quad \cdot \quad \cdot,$$

$$\frac{dx_n}{dt} = a_{n1}(t)x_1 + a_{n2}(t)x_2 + \cdots + a_{nn}(t)x_n.$$

We have taken t (which will often represent *time*) as the independent variable, and the variables $x_1, x_2, \ldots, x_n$ are the unknowns. We shall suppose that the functions $a_{ij}(t)$ are continuous on an interval I. The interval may be any of the types $[a, b]$, (a, b), $[a, b)$, $(a, b]$. In the second and fourth types, a may be $-\infty$, while in the second and third, b may be $+\infty$.

By a *solution* $x(t)$ of such a system is meant a column of functions

(1.2)
$$
\begin{array}{c}
x_1(t) \\
x_2(t) \\
\vdots \\
x_n(t),
\end{array}
$$

each of class C' in I, which simultaneously satisfy (1.1).

The fundamental existence theorem (see Chapter 3) applied to this system leads to the following result.

Theorem 1.1. If $a_1, a_2, \ldots, a_n$ is an arbitrary set of constants and if t_0 is any point of the interval I, there exists one and only one solution of (1.1) with the property that

$$
x_1(t_0) = a_1, \qquad x_2(t_0) = a_2, \qquad \ldots, \qquad x_n(t_0) = a_n.
$$

This solution is defined on the entire interval I.

Corollary. A solution (1.2) of (1.1) with the property that at some point t_0 of I

$$
x_1(t_0) = x_2(t_0) = \cdots = x_n(t_0) = 0
$$

is identically zero on I; that is,

$$
x_1(t) \equiv x_2(t) \equiv \cdots \equiv x_n(t) \equiv 0.
$$

The proof of the corollary is immediate. We note that if $u(t)$ and $v(t)$ are solutions, so is $u(t) + v(t)$, where by $u(t) + v(t)$ is meant the column

$$
\begin{array}{c}
u_1(t) + v_1(t) \\
u_2(t) + v_2(t) \\
\vdots \\
u_n(t) + v_n(t).
\end{array}
$$

So also is $cu(t)$, where by $cu(t)$ is meant the column

$$
\begin{array}{c}
cu_1(t) \\
cu_2(t) \\
\vdots \\
cu_n(t).
\end{array}
$$

It follows that if c and k are constants and $u(t)$ and $v(t)$ are solutions of (1.1), so also is the column $cu(t) + kv(t)$.

It is frequently desirable to transform a simple linear differential equation of order n into a system of the type (1.1). This can be accomplished as follows. Suppose the given differential equation is

$$(1.3) \qquad \frac{d^n x}{dt^n} = a_1(t) \frac{d^{n-1} x}{dt^{n-1}} + a_2(t) \frac{d^{n-2} x}{dt^{n-2}} + \cdots + a_{n-1}(t) \frac{dx}{dt} + a_n(t) x,$$

where the functions $a_i(t)$ are continuous on the interval I. If we set

$$(1.4) \qquad x_1 = x, \qquad x_2 = \frac{dx}{dt}, \qquad x_3 = \frac{d^2 x}{dt^2}, \qquad \cdots, \qquad x_n = \frac{d^{n-1} x}{dt^{n-1}},$$

equation (1.3) is transformed under (1.4) into the system

$$\frac{dx_1}{dt} = x_2,$$

$$\frac{dx_2}{dt} = x_3,$$

$$\cdot \quad \cdot \quad \cdot \quad \cdot$$

$$\frac{dx_{n-1}}{dt} = x_n,$$

$$\frac{dx_n}{dt} = a_1(t) x_n + a_2(t) x_{n-1} + \cdots + a_n(t) x_1.$$

Thus, a system (1.1) may be regarded as a generalization of a single linear differential equation of order n.

For the remainder of this section we shall assume $n = 3$ in order to simplify slightly the presentation of the proofs. The alterations required for the general case will be evident.

We shall, as usual, use dots to denote differentiation with respect to t; thus

$$\dot{x} = \frac{dx}{dt}, \qquad \ddot{x} = \frac{d^2 x}{dt^2}, \qquad \dot{a}(t) = \frac{d}{dt} a(t).$$

When $n = 3$, the system (1.1) becomes

$$\frac{dx_1}{dt} = a_{11}(t) x_1 + a_{12}(t) x_2 + a_{13}(t) x_3,$$

$$(1.1)' \qquad \frac{dx_2}{dt} = a_{21}(t) x_1 + a_{22}(t) x_2 + a_{23}(t) x_3,$$

$$\frac{dx_3}{dt} = a_{31}(t) x_1 + a_{32}(t) x_2 + a_{33}(t) x_3,$$

and a solution $x(t)$ becomes the column

(1.2)′
$$x_1(t)$$
$$x_2(t)$$
$$x_3(t).$$

Three solutions $u(t)$, $v(t)$, $w(t)$ are said to be *linearly dependent* if there exist constants c_1, c_2, c_3, not all zero, such that

(1.5)
$$c_1 u(t) + c_2 v(t) + c_3 w(t) \equiv 0.$$

Condition (1.5) may be interpreted as follows. Let the components of $u(t)$, $v(t)$, $w(t)$ be given, respectively, by the columns

$$u_1(t) \qquad v_1(t) \qquad w_1(t)$$
$$u_2(t) \qquad v_2(t) \qquad w_2(t)$$
$$u_3(t), \qquad v_3(t), \qquad w_3(t).$$

Then (1.5) is equivalent to the three conditions

(1.5)′
$$c_1 u_1(t) + c_2 v_1(t) + c_3 w_1(t) \equiv 0,$$
$$c_1 u_2(t) + c_2 v_2(t) + c_3 w_2(t) \equiv 0,$$
$$c_1 u_3(t) + c_2 v_3(t) + c_3 w_3(t) \equiv 0.$$

Solutions not linearly dependent are said to be *linearly independent*.

Theorem 1.2. There exist three linearly independent solutions of (1.1)′. *Every solution of the system can be written as a linear combination of these three solutions.*

The solutions $u(t)$, $v(t)$, $w(t)$ defined, respectively, by the conditions

(1.6)
$$u_1(t_0) = 1, \qquad v_1(t_0) = 0, \qquad w_1(t_0) = 0,$$
$$u_2(t_0) = 0, \qquad v_2(t_0) = 1, \qquad w_2(t_0) = 0,$$
$$u_3(t_0) = 0, \qquad v_3(t_0) = 0, \qquad w_3(t_0) = 1 \qquad (a \le t_0 \le b),$$

are readily shown to be linearly independent. For, if constants c_1, c_2, c_3, not all zero, exist such that (1.5)′ holds, these conditions would hold in particular when $t = t_0$. But this is impossible. [Note that the solutions $u(t)$, $v(t)$, $w(t)$ defined by (1.6) *do* exist, because of Theorem 1.1.]

It remains to prove that every solution $x(t)$ of (1.1)′ may be written as a linear combination of $u(t)$, $v(t)$, $w(t)$. We shall show that if $x_1(t)$, $x_2(t)$, $x_3(t)$ are the components of $x(t)$, then

$$x(t) \equiv x_1(t_0)u(t) + x_2(t_0)v(t) + x_3(t_0)w(t);$$

that is,

$$x_i(t) \equiv x_1(t_0)u_i(t) + x_2(t_0)v_i(t) + x_3(t_0)w_i(t) \qquad (i = 1, 2, 3).$$

First, note that $X(t)$, with components

$$X_i(t) = x_i(t) - [x_1(t_0)u_i(t) + x_2(t_0)v_i(t) + x_3(t_0)w_i(t)] \quad (i = 1, 2, 3),$$

is a solution of (1.1)′. This solution has the property that

$$X_1(t_0) = X_2(t_0) = X_3(t_0) = 0.$$

Thus, by the corollary to Theorem 1.1, $X_i(t) \equiv 0$. The proof of the theorem is complete.

Exercises

1. Prove that if $u(t)$, $v(t)$, $w(t)$ are any three linearly independent solutions of (1.1)′, every solution of the system may be written as a linear combination of these three solutions.

2. Prove that any four solutions of (1.1)′ are linearly dependent.

3. Given the system of differential equations

$$\frac{dx}{dt} = \frac{1}{2}x + \frac{1}{2}z,$$

$$\frac{dz}{dt} = -\frac{3}{2}x + \frac{5}{2}z,$$

show that each column below provides a solution of the system:

$$
\begin{array}{cc}
e^t & e^{2t} \\
e^t, & 3e^{2t}.
\end{array}
$$

Are the solutions linearly dependent ?

4. Given the system

$$\frac{dx_1}{dt} = \frac{t}{t-1}x_2 + \frac{1}{1-t}x_3,$$

$$\frac{dx_2}{dt} = \frac{1}{t(1-t)}x_2 + \frac{1}{t(t-1)}x_3,$$

$$\frac{dx_3}{dt} = \frac{2}{1-t}x_2 + \frac{2}{t-1}x_3,$$

show that the columns

$$\begin{matrix} 1 & t & 1 \\ 0 & 1 & t \\ 0, & 1, & t^2 \end{matrix}$$

are linearly independent solutions.

5.   Show that the four solutions

$$\begin{matrix} u\text{:} & v\text{:} & w\text{:} & z\text{:} \\ 1 & t & 1 & t+2 \\ 0 & 1 & t & t+1 \\ 0, & 1, & t^2, & t^2+1 \end{matrix}$$

of the system in Exercise 4 are linearly dependent by finding four constants c_1, c_2, c_3, c_4, not all zero, such that

$$c_1 u + c_2 v + c_3 w + c_4 z \equiv 0.$$

2   Fundamental systems of solutions

In this section, as in Section 1, we shall devote our attention to the system (1.1)′ as a representative of the more general system (1.1).

Let the columns

$$\begin{matrix} u_1(t) & v_1(t) & w_1(t) \\ u_2(t) & v_2(t) & w_2(t) \\ u_3(t), & v_3(t), & w_3(t) \end{matrix}$$

be solutions $u(t)$, $v(t)$, and $w(t)$, respectively, of (1.1)′ and form the determinant

$$\Delta(t) = \begin{vmatrix} u_1(t) & v_1(t) & w_1(t) \\ u_2(t) & v_2(t) & w_2(t) \\ u_3(t) & v_3(t) & w_3(t) \end{vmatrix}.$$

Recall that

$$
\dot\Delta(t) = \begin{vmatrix} \dot u_1(t) & \dot v_1(t) & \dot w_1(t) \\ u_2(t) & v_2(t) & w_2(t) \\ u_3(t) & v_3(t) & w_3(t) \end{vmatrix} + \begin{vmatrix} u_1(t) & v_1(t) & w_1(t) \\ \dot u_2(t) & \dot v_2(t) & \dot w_2(t) \\ u_3(t) & v_3(t) & w_3(t) \end{vmatrix}
$$

(2.1)

$$
+ \begin{vmatrix} u_1(t) & v_1(t) & w_1(t) \\ u_2(t) & v_2(t) & w_2(t) \\ \dot u_3(t) & \dot v_3(t) & \dot w_3(t) \end{vmatrix}.
$$

From (1.1)' we see that

$$\dot{u}_i = a_{i1}u_1 + a_{i2}u_2 + a_{i3}u_3,$$

(2.2)
$$\dot{v}_i = a_{i1}v_1 + a_{i2}v_2 + a_{i3}v_3,$$

$$\dot{w}_i = a_{i1}w_1 + a_{i2}w_2 + a_{i3}w_3 \qquad (i = 1, 2, 3).$$

If the dot quantities in the determinants (2.1) are replaced by the equivalent quantities given by (2.2), it will be seen that

(2.3)
$$\dot{\Delta}(t) \equiv [a_{11}(t) + a_{22}(t) + a_{33}(t)]\Delta(t).$$

To verify this, note that the first determinant in (2.1) becomes

$$\Delta_1 = \begin{vmatrix} a_{11}u_1 + a_{12}u_2 + a_{13}u_3 & a_{11}v_1 + a_{12}v_2 + a_{13}v_3 & a_{11}w_1 + a_{12}w_2 + a_{13}w_3 \\ u_2 & v_2 & w_2 \\ u_3 & v_3 & w_3 \end{vmatrix}.$$

If a_{12} times the second row and a_{13} times the third row of this determinant are subtracted term-by-term from the first row, the value of the determinant is unchanged and

$$\Delta_1 = \begin{vmatrix} a_{11}u_1 & a_{11}v_1 & a_{11}w_1 \\ u_2 & v_2 & w_2 \\ u_3 & v_3 & w_3 \end{vmatrix} = a_{11}\Delta(t).$$

In a similar fashion, the second and third determinants in (2.1) can be shown to have the values $a_{22}\Delta(t)$ and $a_{33}\Delta(t)$, respectively. The result (2.3) then follows.

We note from (2.3) that $\Delta(t)$ is a solution of the linear differential equation

$$\dot{x} = [a_{11}(t) + a_{22}(t) + a_{33}(t)]x.$$

It follows that $\Delta(t)$ is either never zero or is identically zero. Indeed,

$$\Delta(t) = \Delta(t_0)e^{\int_{t_0}^{t} [a_{11}(t) + a_{22}(t) + a_{33}(t)]dt},$$

where t_0 is any point of the interval I.

A system of three solutions of (1.1)' with the property that $\Delta(t) \neq 0$ is called a *fundamental system* of solutions.

Theorem 2.1. A necessary and sufficient condition that three solutions of (1.1)′ be linearly independent is that they form a fundamental system.

The proof is left to the student.

Exercises

1.   Prove Theorem 2.1.

2.   Compute $\Delta(t)$ for the system of solutions given for Exercise 3, Section 1, and thus verify that

$$\dot{\Delta}(t) \equiv [a_{11}(t) + a_{22}(t)]\Delta(t).$$

3.   Compute $\Delta(t)$ for the solution given for Exercise 4, Section 1, and thus verify that

$$\dot{\Delta}(t) \equiv [a_{11}(t) + a_{22}(t) + a_{33}(t)]\Delta(t).$$

Does this example invalidate the conclusion of the text that $\Delta(t)$ is either identically zero or never zero?

3   Linear systems with constant coefficients; matrix notation

We shall begin by recalling methods of solving systems of the type

(3.1)

$$\frac{dx_1}{dt} = a_{11}x_1 + a_{12}x_2,$$

$$\frac{dx_2}{dt} = a_{21}x_1 + a_{22}x_2,$$

where a_{11}, a_{12}, a_{21}, and a_{22} are constants. The following examples will illustrate techniques for solving such a system.

Example.   To solve the system

(3.2)

$$\frac{dx_1}{dt} = \frac{1}{2}x_1 + \frac{1}{2}x_2,$$

$$\frac{dx_2}{dt} = -\frac{3}{2}x_1 + \frac{5}{2}x_2$$

let us attempt to determine constants A, B, and λ such that the column

$$Ae^{\lambda t}$$

$$Be^{\lambda t}$$

is a solution of (3.2). If we substitute these functions for x_1 and x_2, respectively, in (3.2), we see that A, B, and λ must satisfy the equations

(3.3)
$$(\tfrac{1}{2} - \lambda)A + \tfrac{1}{2}B = 0,$$

$$-\tfrac{3}{2}A + (\tfrac{5}{2} - \lambda)B = 0.$$

For solutions A and B, not both zero, of equations (3.3) to exist, it is necessary and sufficient that

(3.4)
$$\begin{vmatrix} \tfrac{1}{2} - \lambda & \tfrac{1}{2} \\ -\tfrac{3}{2} & \tfrac{5}{2} - \lambda \end{vmatrix} = 0.$$

Equation (3.4) is readily seen to be equivalent to

$$\lambda^2 - 3\lambda + 2 = 0,$$

the roots of which are 1 and 2. If we set $\lambda = 1$ in (3.3), we obtain

(3.5)
$$-\frac{A}{2} + \frac{B}{2} = 0,$$

$$-\frac{3A}{2} + \frac{3B}{2} = 0.$$

These simultaneous equations are consistent, and we may choose as solutions of them any pair of numbers different from zero satisfying them. An obvious choice is $A = B = 1$. This leads to the solution

$$e^t$$

$$e^t.$$

If we had chosen any other pair of numbers A, B satisfying (3.5), we would simply have obtained a constant times this solution.

In a similar way, the root $\lambda = 2$ leads to the equations

$$-\frac{3A}{2} + \frac{B}{2} = 0,$$

$$-\frac{3A}{2} + \frac{B}{2} = 0.$$

A nontrivial solution of this system is $A = 1$, $B = 3$, and the corresponding solution of (3.2) is

$$e^{2t}$$

$$3e^{2t}.$$

It is easy to verify that these two solutions are linearly independent, and accordingly, every solution of (3.2) may be written in the form

(3.6)
$$c_1 e^t + c_2 e^{2t},$$

$$c_1 e^t + 3c_2 e^{2t},$$

where c_1 and c_2 are constants. Since, conversely, for every choice of the constants c_1 and c_2 the column (3.6) is a solution, it is appropriate to refer to (3.6) as the *general solution* of (3.2).

Example.   Solve the system

(3.7)
$$\frac{dx_1}{dt} = 3x_1 - x_2,$$

$$\frac{dx_2}{dt} = x_1 + x_2.$$

The substitution in (3.7) of the column

$$Ae^{\lambda t}$$

$$Be^{\lambda t}$$

leads in this case to the equations

(3.8)
$$(3 - \lambda)A - \qquad B = 0,$$

$$A + (1 - \lambda)B = 0,$$

and the determinantal equation

(3.9)
$$\begin{vmatrix} 3 - \lambda & -1 \\ 1 & 1 - \lambda \end{vmatrix} = 0.$$

The roots of (3.9) are seen to be $\lambda = 2, 2$. Setting $\lambda = 2$ in (3.8), we determine the nontrivial solution $A = B = 1$ and the corresponding solution

(3.10)
$$e^{2t}$$

$$e^{2t}$$

of the system (3.7). Since the roots of (3.9) are equal, we cannot employ the second root of this quadratic equation to obtain a second linearly independent solution of (3.7). The situation here is similar to that which arose when we solved a linear differential equation of second order whose corresponding indicial equation possessed a repeated root (see Chapter 2, Section 2). This similarity might suggest that a second solution of (3.7) could be obtained by multiplying the solution (3.10) by t:

$$te^{2t}$$

$$te^{2t}.$$

It is easy to see, however, by testing these functions in (3.7), that they do not provide a solution. A more promising scheme is to attempt to determine functions $u(t)$ and $v(t)$ such that

$$e^{2t}u(t)$$

$$e^{2t}v(t)$$

is a solution. Upon substituting these functions for x_1 and x_2, respectively, in (3.7), we find

$$\dot{u} = u - v,$$

$$\dot{v} = u - v.$$

It follows that $\dot{u} = \dot{v}$, and hence that $u = v + c_1$. Thus, $c_1 = u - v = \dot{u} = \dot{v}$; hence, $u = c_1 t + c_2$, $v = c_1 t + (c_2 - c_1)$. We choose $c_2 = 0$, $c_1 = 1$, and we have the solution

$$te^{2t}$$

(3.11)

$$(t - 1)e^{2t}$$

of (3.7). The solutions (3.10) and (3.11) may be proved to be linearly independent by computing the determinant

$$\Delta(t) = \begin{vmatrix} e^{2t} & te^{2t} \\ e^{2t} & (t-1)e^{2t} \end{vmatrix} = -e^{4t}.$$

Since $\Delta(t) \neq 0$, the solutions are linearly independent.

The general solution of equations (3.7) is, accordingly,

$$c_1 e^{2t} + c_2 te^{2t}$$

$$c_1 e^{2t} + c_2 (t - 1)e^{2t}.$$

A somewhat shorter procedure after obtaining the solution (3.10) would be to determine constants a, b, f, g such that

$$(at + b)e^{2t}$$

$$(ft + g)e^{2t}$$

is a solution. This can be accomplished by substituting these functions in (3.7).

A little algebra will show that when the characteristic roots are equal and $a_{22} \neq a_{11}$ in (3.1), we may take the second solution as

$$t(Ae^{\lambda_1 t})$$

$$(t + \alpha)(Be^{\lambda_1 t}),$$

where λ_1 is the repeated root of the *characteristic equation*

$$\begin{vmatrix} a_{11} - \lambda & a_{12} \\ a_{21} & a_{22} - \lambda \end{vmatrix} = 0,$$

and

$$\alpha = \frac{2}{a_{22} - a_{11}}.$$

In the present example, $a_{11} = 3$, $a_{22} = 1$, and $\alpha = -1$.

When $a_{22} = a_{11}$ in (3.1), and the roots λ are equal, the method given above breaks down. In this case, either $a_{12} = 0$ or $a_{21} = 0$, or both, for the characteristic equation becomes

$$\begin{vmatrix} a_{11} - \lambda & a_{12} \\ a_{21} & a_{11} - \lambda \end{vmatrix} = 0,$$

or

$$(\lambda - a_{11})^2 = a_{12}a_{21}.$$

We note that the roots of this equation are then $\lambda = a_{11}, a_{11}$. The system (3.1) becomes either

$$\frac{dx_1}{dt} = a_{11}x_1,$$

$$\frac{dx_2}{dt} = a_{21}x_1 + a_{11}x_2,$$

or

$$\frac{dx_1}{dt} = a_{11}x_1 + a_{12}x_2,$$

$$\frac{dx_2}{dt} = a_{11}x_2.$$

The treatments of these systems are similar, so we shall examine only the former. The general solution of the first equation is

$$x_1 = c_1 e^{a_{11} t} \qquad (c_1 \text{ constant}).$$

We substitute this value in the second equation and we have

$$\frac{dx_2}{dt} - a_{11}x_2 = c_1 a_{21} e^{a_{11}t},$$

with the general solution

$$x_2 = c_1 a_{21} t e^{a_{11}t} + c_2 e^{a_{11}t}.$$

Thus, linearly independent solutions of the system are

$$0 \qquad e^{a_{11}t}$$
$$e^{a_{11}t}, \quad a_{21} t e^{a_{11}t}.$$

A similar treatment of the second system yields the linearly independent solutions

$$e^{a_{11}t} \quad a_{12} t e^{a_{11}t}$$
$$0, \qquad e^{a_{11}t}.$$

Note that when $a_{11} = 0$, the last pair of solutions becomes

$$1 \quad a_{12}t$$
$$0, \quad 1.$$

Example. Solve the system

$$\frac{dx_1}{dt} = 3x_1 + 2x_2,$$

(3.12)

$$\frac{dx_2}{dt} = -x_1 + x_2.$$

The substitution in (3.12) of the column

$$Ae^{\lambda t}$$
$$Be^{\lambda t}$$

leads to the equations

(3.13)
$$(3 - \lambda)A + 2B = 0,$$
$$-A + (1 - \lambda)B = 0,$$

and the determinantal equation

(3.14)
$$\begin{vmatrix} 3 - \lambda & 2 \\ -1 & 1 - \lambda \end{vmatrix} = 0.$$

The roots λ of (3.14) are $2 + i$ and $2 - i$. The substitution of $\lambda = 2 + i$ in (3.13) leads to the nontrivial solution $A = 1 + i$, $B = -1$ of (3.13) and the solution

$$(1 + i)e^{(2+i)t}$$

(3.15)

$$-e^{(2+i)t}$$

of (3.12). The solution (3.15) can be rewritten, using the familiar relationship (see Chapter 2, Section 3)

$$e^{(a+ib)t} = e^{at}(\cos bt + i \sin bt),$$

in the form

$$e^{2t}[(\cos t - \sin t) + i(\cos t + \sin t)]$$

$$-e^{2t}[\cos t + i \sin t].$$

The real part of this solution,

(3.16)
$$e^{2t}(\cos t - \sin t)$$

$$-e^{2t}\cos t,$$

must then also be a solution. So also is the imaginary part

(3.17)
$$e^{2t}(\cos t + \sin t)$$

$$-e^{2t}\sin t.$$

The solutions (3.16) and (3.17) may, of course, be tested directly in (3.12). The computation of

$$\Delta(t) = \begin{vmatrix} e^{2t}(\cos t - \sin t) & e^{2t}(\cos t + \sin t) \\ -e^{2t}\cos t & -e^{2t}\sin t \end{vmatrix} = e^{4t}$$

shows that the solutions (3.16) and (3.17) are linearly independent.

The general solution of (3.12) is, accordingly,

$$c_1 e^{2t}(\cos t - \sin t) + c_2 e^{2t}(\cos t + \sin t)$$

$$-c_1 e^{2t}\cos t - c_2 e^{2t}\sin t.$$

Since the root $\lambda = 2 + i$ of (3.14) yielded the general solution of (3.12), we can gain nothing by the substitution of the second root $\lambda = 2 - i$ of equation (3.14).

The three examples above illustrate all the possibilities for a system (3.1) since the roots λ of the corresponding determinantal equation

$$\begin{vmatrix} a_{11} - \lambda & a_{12} \\ a_{21} & a_{22} - \lambda \end{vmatrix} = 0$$

fall into one of the three following categories:

1. Roots real and distinct.
2. Roots real and equal.
3. Roots conjugate complex.

The substitution of

$$Ae^{\lambda t}$$
$$Be^{\lambda t}$$
$$Ce^{\lambda t}$$

in a system of three linear differential equations with constant coefficients leads to similar analysis. The student should experience no difficulty in carrying out the details.

Nonhomogeneous equations. Consider the *nonhomogeneous* system

(3.18)
$$\dot{x}_1 = \tfrac{1}{2}x_1 + \tfrac{1}{2}x_2 + t - 1,$$
$$\dot{x}_2 = -\tfrac{3}{2}x_1 + \tfrac{5}{2}x_2 - 2t.$$

The *corresponding homogeneous* system [see (3.2)]

$$\dot{x}_1 = \tfrac{1}{2}x_1 + \tfrac{1}{2}x_2,$$
$$\dot{x}_2 = -\tfrac{3}{2}x_1 + \tfrac{5}{2}x_2$$

has the general solution

(3.19)
$$x_1 = c_1 e^t + c_2 e^{2t},$$
$$x_2 = c_1 e^t + 3c_2 e^{2t}.$$

We seek a particular solution of the system (3.18). To find it, we try to determine constants a, b, h, and k so that the column

(3.20)
$$x_1 = at + b,$$
$$x_2 = ht + k,$$

is a solution. Substituting (3.20) in (3.18) we find that

$$a = -\tfrac{7}{4}, \qquad b = -\tfrac{7}{8}, \qquad h = -\tfrac{1}{4}, \qquad k = -\tfrac{5}{8},$$

and a particular solution of (3.18) is then

(3.21)
$$x_1 = -\tfrac{7}{4}t - \tfrac{7}{8},$$
$$x_2 = -\tfrac{1}{4}t - \tfrac{5}{8}.$$

The treatment of the nonhomogeneous single linear differential equation (Chapter 2, Section 4) suggests that the general solution of (3.18) ought to be the "sum" of (3.21) and (3.19); that is,

$$x_1 = -\tfrac{7}{4}t - \tfrac{7}{8} + c_1 e^t + c_2 e^{2t},$$

$$x_2 = -\tfrac{1}{4}t - \tfrac{5}{8} + c_1 e^t + 3c_2 e^{2t}.$$

This is the case, as will be noted in the general result in Theorem 3.1 below.

On vector and matrix notation.   We may consider the column

$$x = \begin{bmatrix} x_1 \\ x_2 \\ \vdots \\ x_n \end{bmatrix}$$

as an n-dimensional vector the kth component of which is x_k. It may also be regarded as an $n \times 1$ matrix. If the components are functions of t on I, the column becomes a *vector-valued* function on I, since each value of t on the interval determines a vector x. The vector-valued function is said to be continuous on I if each component is continuous on I.

The derivative with respect to t of the vector x is the vector

$$\dot{x} = \begin{bmatrix} \dot{x}_1 \\ \dot{x}_2 \\ \vdots \\ \dot{x}_n \end{bmatrix}.$$

The sum of two vectors x and y is the vector

$$x + y = \begin{bmatrix} x_1 + y_1 \\ x_2 + y_2 \\ \vdots \\ x_n + y_n \end{bmatrix},$$

where y_k ($k = 1, 2, \ldots, n$) is, of course, the kth component of y. The integral of a vector-valued function $x(t)$ is the vector

$$\int_a^b x(t)\, dt = \begin{bmatrix} \int_a^b x_1(t)\, dt \\ \int_a^b x_2(t)\, dt \\ \vdots \\ \int_a^b x_n(t)\, dt \end{bmatrix}.$$

If c is any real number, the vector cx is the vector whose kth component is cx_k. Two vectors x and y are said to be equal if and only if $x_k = y_k$ ($k = 1, 2, \ldots, n$). The *null vector* 0 is the vector each of whose components is zero. The *norm* $\|x\|$ of a vector x may be defined in a number of ways. The usual definition is

$$\|x\| = (x_1^2 + x_2^2 + \cdots + x_n^2)^{1/2}.$$

If we designate by $A(t)$ the matrix

$$A(t) = \begin{pmatrix} a_{11}(t) & a_{12}(t) & \cdots & a_{1n}(t) \\ a_{21}(t) & a_{22}(t) & \cdots & a_{2n}(t) \\ \cdot & \cdot & \cdots & \cdot \\ a_{n1}(t) & a_{n2}(t) & \cdots & a_{nn}(t) \end{pmatrix},$$

the product $A(t)x$ of the matrices $A(t)$ and x is precisely the right-hand member of (1.1), and the system (1.1) can then be written

(3.22) $$\frac{dx}{dt} = A(t)x.$$

We say that $A(t)$ is a *continuous* matrix if each element $a_{ij}(t)$ is continuous. A solution of (3.22) is then a vector-valued function $x(t)$—that is, a column such as (1.2).

Theorem 1.1 can now be translated as follows:

Suppose $A(t)$ is continuous on $I: a \leq t \leq b$, let t_0 be any point of I, and let

$$a = \begin{bmatrix} a_1 \\ a_2 \\ \vdots \\ a_n \end{bmatrix}$$

be an arbitrary constant vector (or $n \times 1$ matrix). There then exists one and *only one solution $x(t)$ of the (matrix) differential system*

$$\frac{dx}{dt} = A(t)x,$$

$$x(t_0) = a.$$

In what follows we shall employ only the most elementary operations with matrices and vectors.

Consider now the general linear system

(3.23) $$\dot{x} = A(t)x + f(t),$$

where x, $\dot{x}$, $f(t)$ are (column) n-vectors, $f(t)$ is continuous, and $A(t)$ is an $n \times n$ continuous matrix:

Theorem 3.1. If $x^0(t)$ is any solution of (3.23), the general solution is given by

$$x^0(t) + X(t),$$

where $X(t)$ is the general solution of the corresponding homogeneous system

(3.24) $$\dot{x} = A(t)x.$$

Equation (3.24) will be seen to be equations (1.1) written in matrix-vector form.

To prove the theorem, let $x^*(t)$ be an arbitrary solution of (3.23). Then

$$\dot{x}^*(t) \equiv A(t)x^*(t) + f(t),$$
$$\dot{x}^0(t) \equiv A(t)x^0(t) + f(t),$$

and, consequently,

$$[x^*(t) - x^0(t)]^{\cdot} \equiv A(t)[x^*(t) - x^0(t)];$$

that is, $x^*(t) - x^0(t)$ is a solution of the system (3.24). It follows that

$$x^*(t) \equiv x^0(t) + X(t),$$

where $X(t)$ is a solution of (3.24).

Conversely, if $X(t)$ is an arbitrary solution of (3.24) and $x^0(t)$ is any particular solution of (3.23), the sum

$$x^0(t) + X(t)$$

is also a solution of (3.23), as is readily verified.

The proof of the theorem is complete.

Variation of parameters. It is natural to inquire whether or not a particular solution of (3.23) can be found if a fundamental solution of the corresponding homogeneous equation (3.24) is known. The answer is provided by the following analysis.

Let $X(t)$ be a fundamental solution of the homogeneous differential equation

(3.25) $$\dot{x} = A(t)x,$$

and consider the substitution

(3.26) $$x = X(t)z$$

in the equation

(3.27) $$\dot{x} = A(t)x + f(t).$$

Differentiating (3.26), we have

$$\dot{x} = \dot{X}z + X\dot{z},$$

and thus

$$\dot{X}z + X\dot{z} = AXz + f,$$

or

$$AXz + X\dot{z} = AXz + f.$$

Inasmuch as $X(t)$ is nonsingular, its inverse $X^{-1}(t)$ exists, and we have

$$\dot{z} = X^{-1}(t)f.$$

It follows that

$$z(t) = c + \int_{t_0}^{t} X^{-1}(t)f(t)\, dt \qquad (t_0 \in I),$$

where c is a constant vector.

Thus a solution of (3.27) may be written

(3.28) $$x(t) = X(t)\left[c + \int_{t_0}^{t} X^{-1}(t)f(t)\, dt \right],$$

where $x(t_0) = X(t_0)c$.

Example. Consider the equation

$$\dot{x} = A(t)x + f(t),$$

where

$$A(t) = \begin{Vmatrix} 0 & 1 \\ -2 & -3 \end{Vmatrix}, \qquad f(t) = \begin{Vmatrix} t \\ -t^2 \end{Vmatrix};$$

that is, consider the system

$$\dot{x} = y + t,$$
$$\dot{y} = -2x - 3y - t^2.$$

A fundamental system of solutions of the homogeneous equation $\dot{x} = Ax$ is readily seen to be

$$X(t) = \begin{Vmatrix} e^{-t} & e^{-2t} \\ -e^{-t} & -2e^{-2t} \end{Vmatrix}.$$

Let the inverse of $X(t)$ be given by

$$X^{-1}(t) = \begin{Vmatrix} \alpha & \beta \\ \gamma & \delta \end{Vmatrix}.$$

Then using the fact that

$$X(t)X^{-1}(t) = \begin{Vmatrix} 1 & 0 \\ 0 & 1 \end{Vmatrix},$$

we find that

$$\alpha = 2e^t, \qquad \beta = e^t,$$
$$\gamma = -e^{2t}, \qquad \delta = -e^{2t}.$$

Equation (3.28) yields, successively,

$$x(t) = X(t)c + X(t) \int_0^t \begin{Vmatrix} 2\,e^t & e^t \\ -e^{2t} & -e^{2t} \end{Vmatrix} \cdot \begin{Vmatrix} t \\ -t^2 \end{Vmatrix} dt$$

$$= X(t)c + X(t) \int_0^t \begin{Vmatrix} 2\,te^t - t^2e^t \\ -te^{2t} + t^2e^{2t} \end{Vmatrix} dt$$

$$= X(t)c + \begin{Vmatrix} e^{-t} & e^{-2t} \\ -e^{-t} & -2e^{-2t} \end{Vmatrix} \cdot \begin{Vmatrix} -t^2e^t + 4te^t - 4e^t + 4 \\ \frac{1}{2}t^2e^{2t} - te^{2t} + \frac{1}{2}e^{2t} - \frac{1}{2} \end{Vmatrix}$$

$$= X(t)c + \left\| \begin{array}{c} -\dfrac{t^2}{2} + 3t - \dfrac{7}{2} + 4e^{-t} - \dfrac{1}{2}e^{-2t} \\[2mm] -2t + 3 - 4e^{-t} + e^{-2t} \end{array} \right\|.$$

This is a particular solution of (3.27) for each choice of a constant vector c. If we choose

$$c = \left\| \begin{array}{c} -4 \\ \frac{1}{2} \end{array} \right\|,$$

we obtain the particular solution

$$x(t) = \left\| \begin{array}{c} -\dfrac{t^2}{2} + 3t - \dfrac{7}{2} \\[2mm] -2t + 3 \end{array} \right\|.$$

Exercises

Find the general solutions of the given systems of differential equations.

1. $\dfrac{dx_1}{dt} = x_2,$ $\dfrac{dx_2}{dt} = -2x_1 + 3x_2.$

2. $\dfrac{dx_1}{dt} = -3x_1 + 4x_2,$ $\dfrac{dx_2}{dt} = -2x_1 + 3x_2.$

3. $\dfrac{dx_1}{dt} = x_2,$ $\dfrac{dx_2}{dt} = -x_1 + 2x_2.$

4. $\dfrac{dx_1}{dt} = 4x_1 - x_2,$ $\dfrac{dx_2}{dt} = 4x_1.$

5. $\dfrac{dx_1}{dt} = -2x_2,$ $\dfrac{dx_2}{dt} = x_1 + 2x_2.$

6. $\dfrac{dx_1}{dt} = x_1 - 2x_2,$ $\dfrac{dx_2}{dt} = 4x_1 + 5x_2.$

7. $\dfrac{dx_1}{dt} = 2x_1 + x_2 - 2x_3,$

    $\dfrac{dx_2}{dt} = 3x_2 - 2x_3,$

    $\dfrac{dx_3}{dt} = 3x_1 + x_2 - 3x_3.$

8. $\dfrac{dx_1}{dt} = x_2,$

   $2\dfrac{dx_2}{dt} = -3x_1 + 6x_2 - x_3,$

   $\dfrac{dx_3}{dt} = -x_1 + x_2 + x_3.$

9. $\dfrac{dx_1}{dt} = 3x_1 - 3x_2 + x_3,$

   $\dfrac{dx_2}{dt} = 2x_1 - x_2,$

   $\dfrac{dx_3}{dt} = x_1 - x_2 + x_3.$

   Find a fundamental system of solutions for each of the following systems.

10. $\dfrac{dx_1}{dt} = x_2,$

   $\dfrac{dx_2}{dt} = x_3,$

   $\dfrac{dx_3}{dt} = 8x_1 - 12x_2 + 6x_3.$

11. $\dfrac{dx_1}{dt} = x_2,$

   $\dfrac{dx_2}{dt} = x_3,$

   $\dfrac{dx_3}{dt} = x_4,$

   $\dfrac{dx_4}{dt} = -2x_1 + x_2 + 3x_3 - x_4.$

12. $\dfrac{dx_1}{dt} = x_2,$

   $\dfrac{dx_2}{dt} = x_3,$

   $\dfrac{dx_3}{dt} = x_4,$

$$\frac{dx_4}{dt} = -x_3 + 2x_4.$$

13. $$\frac{dx_1}{dt} = x_2,$$

$$\frac{dx_2}{dt} = x_3,$$

$$\frac{dx_3}{dt} = x_4,$$

$$\frac{dx_4}{dt} = -x_1 + 2x_2 - 2x_3 + 2x_4.$$

14. Refer to Exercise 1 and solve the system

$$\dot{x}_1 = x_2 + t,$$
$$\dot{x}_2 = -2x_1 + 3x_2 - 1.$$

15. Refer to Exercise 2 and solve the system

$$\dot{x}_1 = -3x_1 + 4x_2 + \sin t,$$
$$\dot{x}_2 = -2x_1 + 3x_2 - 2.$$

16. Refer to Exercise 3 and solve the system

$$\dot{x}_1 = x_2 + 2t,$$
$$\dot{x}_2 = -x_1 + 2x_2 - 3.$$

17. Refer to Exercise 4 and solve the system

$$\dot{x}_1 = 4x_1 - x_2 - 3e^{2t},$$
$$\dot{x}_2 = 4x_1 + 1.$$

18. Solve the matrix differential system

$$\frac{dx}{dt} = A(t)x$$

when

(a) $A(t) = \begin{pmatrix} \frac{1}{2} & \frac{1}{2} \\ -\frac{3}{2} & \frac{5}{2} \end{pmatrix}$, $x(0) = \begin{bmatrix} 1 \\ 1 \end{bmatrix}$;

(b) $A(t) = \begin{pmatrix} 0 & 1 \\ -2 & 3 \end{pmatrix}$, $x(0) = \begin{bmatrix} 0 \\ 1 \end{bmatrix}$;

$$(c) \ A(t) = \begin{pmatrix} 0 & 2 & 0 \\ -1 & 3 & 0 \\ 0 & 0 & 3 \end{pmatrix}, \qquad x(0) = \begin{bmatrix} 0 \\ 1 \\ 1 \end{bmatrix}.$$

19.   Use equation (3.28) to solve the system (3.18). Assume (3.19) as known.

20.   Use equation (3.28) to solve
(a) Exercise 14;
(b) Exercise 15;
(c) Exercise 16;
(d) Exercise 17.

21.   Find the general solution of the system

$$\frac{dx_1}{dt} = x_1 - x_2,$$

$$\frac{dx_2}{dt} = x_2.$$

22.   Find the general solution of the system

$$\frac{dx_1}{dt} = -2x_1,$$

$$\frac{dx_2}{dt} = 3x_1 - 2x_2.$$

23.   Find two linearly independent solutions of the system

$$\frac{dx_1}{dt} = 3x_1,$$

$$\frac{dx_2}{dt} = 3x_2.$$

24.   Find two linearly independent solutions of the system

$$\frac{dx_1}{dt} = 0,$$

$$\frac{dx_2}{dt} = 0.$$

Answers

1. $c_1 e^t + c_2 e^{2t}$,
$c_1 e^t + 2c_2 e^{2t}$.

2. $c_1 e^t + 2c_2 e^{-t}$,
$c_1 e^t + c_2 e^{-t}$.

3. $c_1 e^t + c_2 t e^t,$
    $c_1 e^t + c_2 (t + 1)e^t.$

4. $c_1 e^{2t} + c_2 t e^{2t},$
    $2c_1 e^{2t} + c_2 (2t - 1)e^{2t}.$

5. $2c_1 e^t \cos t + 2c_2 e^t \sin t,$
    $c_1 e^t (\sin t - \cos t) - c_2 e^t (\sin t + \cos t).$

6. $c_1 e^{3t} \cos 2t + c_2 e^{3t} \sin 2t,$
    $c_1 e^{3t} (\sin 2t - \cos 2t) - c_2 e^{3t} (\sin 2t + \cos 2t).$

7. $c_1 e^t + c_2 e^{-t} + c_3 e^{2t},$
    $c_1 e^t + c_2 e^{-t} + 2c_3 e^{2t},$
    $c_1 e^t + 2c_2 e^{-t} + c_3 e^{2t}.$

8. $c_1 e^t + c_2 t e^t + c_3 e^{2t},$
    $c_1 e^t + c_2 (1 + t)e^t + 2c_3 e^{2t},$
    $c_1 e^t + c_2 (2 + t)e^t + c_3 e^{2t}.$

9. $c_1 e^t + c_2 e^t (2 \cos t - \sin t) + c_3 e^t (\cos t + 2 \sin t)$
    $c_1 e^t + 2c_2 e^t \cos t + 2c_3 e^t \sin t$
    $c_1 e^t + c_2 e^t \cos t + c_3 e^t \sin t.$

10. $\begin{matrix} e^{2t} & te^{2t} & t^2 e^{2t} \\ 2e^{2t} & (1 + 2t)e^{2t} & (2t^2 + 2t)e^{2t} \\ 4e^{2t} & (4 + 4t)e^{2t} & (4t^2 + 8t + 2)e^{2t}. \end{matrix}$

11. $\begin{matrix} e^t & e^{-t} & e^{-2t} & te^t \\ e^t & -e^{-t} & -2e^{-2t} & (t + 1)e^t \\ e^t & e^{-t} & 4e^{-2t} & (t + 2)e^t \\ e^t & -e^{-t} & -8e^{-2t} & (t + 3)e^t. \end{matrix}$

12. $\begin{matrix} 1 & e^t & t & te^t \\ 0 & e^t & 1 & (t + 1)e^t \\ 0 & e^t & 0 & (t + 2)e^t \\ 0 & e^t & 0 & (t + 3)e^t. \end{matrix}$

13. $\begin{matrix} e^t & te^t & \cos t & \sin t \\ e^t & (t + 1)e^t & -\sin t & \cos t \\ e^t & (t + 2)e^t & -\cos t & -\sin t \\ e^t & (t + 3)e^t & \sin t & -\cos t. \end{matrix}$

14. $-\frac{3}{2}t - \frac{9}{4} + c_1 e^t + c_2 e^{2t}$
    $-t - \frac{3}{2} + c_1 e^t + 2c_2 e^{2t}.$

15. A particular solution is

$$\tfrac{3}{2} \sin t - \tfrac{1}{2} \cos t + 8,$$

$$\sin t + 6.$$

16.   A particular solution is

$$-4t - 9$$
$$-2t - 4.$$

17.   A particular solution is

$$-3t^2 e^{2t} - 3te^{2t} - \tfrac{1}{4}$$
$$-6t^2 e^{2t} - 1.$$

18.   (a) $x = \begin{bmatrix} e^t \\ e^t \end{bmatrix}$;

(b) $x = \begin{bmatrix} e^{2t} - e^t \\ 2e^{2t} - e^t \end{bmatrix}$;

(c) $x = \begin{bmatrix} 2e^{2t} - 2e^t \\ 2e^{2t} - e^t \\ e^{3t} \end{bmatrix}$.

8

Autonomous systems in the plane

1 The Poincaré phase plane

In a vector differential equation

$$\dot{x} = F(x, t),$$

the right-hand member may not depend formally on the time t. The equation then has the form

$$\dot{x} = f(x).$$

Such a system is said to be *autonomous* or *time-invariant*.

In this chapter we shall be principally concerned with autonomous linear systems

$$\frac{dx}{dt} = ax + by,$$

(1.1)

$$\frac{dy}{dt} = cx + dy,$$

where a, b, c, and d are constants and t is time. Such systems are of funda-mental importance in the study of second-order autonomous systems that are not necessarily linear. We regard t as a parameter and study the curves in the xy-plane—called *trajectories*—defined by equations (1.1). The xy-plane is then said to be the *phase plane* for the system (1.1).

A point at which both dx/dt and dy/dt vanish—where

$$\left(\frac{dx}{dt}\right)^2 + \left(\frac{dy}{dt}\right)^2 = 0$$

—is called an *equilibrium point* of the system (1.1). The origin is such a point. Note that if the determinant

$$\begin{vmatrix} a & b \\ c & d \end{vmatrix} = ad - bc$$

is not equal to 0, the origin is the only equilibrium point of the system (1.1).

We seek a solution of (1.1) in the form

(1.2)
$$x = Ae^{\lambda t},$$
$$y = Be^{\lambda t},$$

where A, B, and λ are suitably chosen constants (see Chapter 7). When the substitution (1.2) is made in equations (1.1), we have

(1.3)
$$A(a - \lambda) + Bb = 0,$$
$$Ac + B(d - \lambda) = 0.$$

Equations (1.3) will possess solutions A and B not both zero if and only if λ is a root of the characteristic equation

(1.4)
$$\begin{vmatrix} a - \lambda & b \\ c & d - \lambda \end{vmatrix} = 0,$$

or

(1.5)
$$\lambda^2 - (a + d)\lambda + (ad - bc) = 0.$$

If equation (1.5) is written as

(1.6)
$$\lambda^2 - p\lambda + q = 0,$$

where

$$p = a + d,$$
$$q = ad - bc,$$

we note that

$$p = \lambda_1 + \lambda_2,$$
$$q = \lambda_1\lambda_2,$$

where λ_1 and λ_2 are the characteristic roots. The discriminant Δ of equation (1.6) is given by

$$\Delta = p^2 - 4q$$
$$= (a - d)^2 + 4bc.$$

The behavior of the solutions of equations (1.1) depends in a fundamental way on the nature of the roots of equation (1.6). Let us examine these solutions more carefully.

If λ is set equal to λ_1 in the first of equations (1.3) and if we designate by A_1 and B_1, respectively, any corresponding nontrivial solution A and B of (1.3), we have

$$A_1(a - \lambda_1) + B_1 b = 0.$$

Similarly,

$$A_2(a - \lambda_2) + B_2 b = 0,$$

where A_2 and B_2 are any nontrivial solution A and B of (1.3) corresponding to the characteristic root λ_2.

If $\lambda_1 \neq \lambda_2$, linearly independent solutions of (1.1) are furnished by the columns

(1.7)
$$A_1 e^{\lambda_1 t} \quad A_2 e^{\lambda_2 t}$$
$$B_1 e^{\lambda_1 t} \quad B_2 e^{\lambda_2 t}.$$

The general solution of (1.1) may then be written in the form

(1.8)
$$x = c_1 A_1 e^{\lambda_1 t} + c_2 A_2 e^{\lambda_2 t},$$
$$y = c_1 B_1 e^{\lambda_1 t} + c_2 B_2 e^{\lambda_2 t},$$

where c_1 and c_2 are arbitrary constants.

The curves in the phase plane (xy-plane) defined by (1.8) satisfy at each point the differential equation

(1.9) $(cx + dy)\, dx = (ax + by)\, dy.$

We observe that, by the fundamental existence theorem, there is one and only one curve through each point of the plane, except possibly the origin.

A number of cases are illustrated below by examples. An examination of the examples will indicate that the methods employed can be applied in general.

Case I. $q < 0$. In this case the characteristic roots λ_1 and λ_2 are real and have opposite signs. As an example, consider the system

$$\frac{dx}{dt} = x - y,$$

(1.10)

$$\frac{dy}{dt} = -2x.$$

The characteristic equation is

$$\begin{vmatrix} 1 - \lambda & -1 \\ -2 & -\lambda \end{vmatrix} = \lambda^2 - \lambda - 2 = 0,$$

and the characteristic roots are $\lambda_1 = -1$, $\lambda_2 = 2$. The general solution may be written in the form

(1.11)

$$x = c_1 e^{-t} + c_2 e^{2t},$$

$$y = 2c_1 e^{-t} - c_2 e^{2t}.$$

Consider first the case when $c_1 = c_2 = 0$. The solution (1.11) becomes the null solution represented in the phase plane by the point at the origin. Next, suppose that $c_2 = 0$ and $c_1 \neq 0$. The points (x, y) of a solution in the phase plane lie on two rays—the line $y = 2x$ with the origin deleted. As t varies from an arbitrary fixed value t_0 to $+\infty$, the point (x, y) moves along its ray toward the origin as a limit. Similarly, if $c_1 = 0$, $c_2 \neq 0$, the point (x, y) moves along the line $y = -x$ (with the origin deleted) away from the origin, as t varies from a fixed value t_0 to $+\infty$.

Either by solving the differential equation (1.9) or by eliminating the parameter t in equation (1.11) we find that the curves represented by (1.11) have the equation

(1.12) $$(x + y)^2(y - 2x) = k,$$

where k is a constant. It is clear from equation (1.12) that the lines $x + y = 0$ and $y - 2x = 0$ are asymptotes of the family (1.12). This can also be observed from equations (1.11). For, when t is very large and positive, the first terms in the right-hand members of (1.11) are very small, and these equations are approximately $x = c_2 e^{2t}$, $y = -c_2 e^{2t}$; that is, the point (x, y) is close to the line $y = -x$. Similarly, when t is numerically large but negative, the point (x, y) is near the line $y = 2x$.

Note that the curves (1.11) have slopes given by the formula

$$\frac{dy}{dx} = \frac{2x}{y - x}.$$

This follows from (1.10). Thus, the *trajectories*—the curves (1.11)—have the property that all have horizontal tangents where they cross the y-axis. Further, they have vertical tangents at the points where they cross the line

$y = x$, and all have slope -2 at their x-intercepts. It is clear from (1.11) that if neither c_1 nor c_2 is zero, a point (x, y) moving along a trajectory cannot approach the origin.

Typical trajectories in the phase plane are shown in Fig. 8.1. The arrowheads on the curves indicate the direction of motion of a point (x, y) along the trajectories, as t increases.

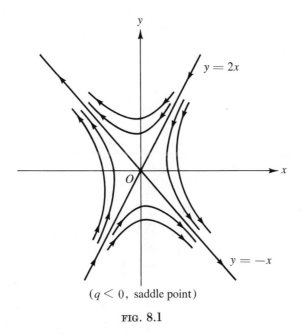

($q < 0$, saddle point)

FIG. 8.1

The equilibrium point at the origin in this case ($q < 0$) is called a *saddle point*.

The notion of the *stability* of the origin—that is, of the nature of motions along the trajectories in the neighborhood of the origin—will be discussed in Chapter 9. For the present we shall say that the origin is *asymptotically stable* when, as in Fig. 8.2, all motions along trajectories approach the origin. When, however, in every neighborhood of the origin there is at least one point with the property that the trajectory through that point does not remain bounded, the origin will be said to be *unstable*. The origin in Fig. 8.1 is unstable.

These definitions apply to linear systems with constant coefficients having the origin as an isolated equilibrium point, but they are consistent with the more general definitions that will be given in Chapter 9.

Case II. $q > 0$, $\Delta > 0$. In this case the characteristic roots λ_1 and λ_2 are real, unequal, and are either both positive or both negative. Consider

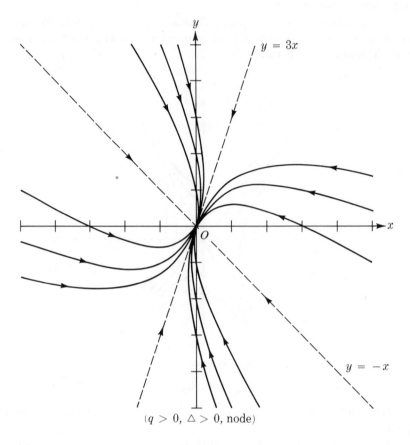

$$(q > 0, \triangle > 0, \text{node})$$

FIG. 8.2

the system

(1.13)

$$\frac{dx}{dt} = -\frac{7}{4}x + \frac{1}{4}y,$$

$$\frac{dy}{dt} = \frac{3}{4}x - \frac{5}{4}y.$$

The characteristic equation is

$$\lambda^2 + 3\lambda + 2 = 0,$$

and the characteristic roots are $\lambda_1 = -1$, $\lambda_2 = -2$. The general solution of (1.13) can be written

$$x = c_1 e^{-t} + c_2 e^{-2t},$$
(1.14)
$$y = 3c_1 e^{-t} - c_2 e^{-2t}.$$

Again, if $c_1 = c_2 = 0$, we obtain the null solution at the origin. If $c_1 = 0$, $c_2 \neq 0$, as t varies from t_0 to $+\infty$, the point (x, y) moves toward the origin along one of the two rays that form the line $y = -x$ with the origin deleted. Similarly, if $c_1 \neq 0$, $c_2 = 0$, as t varies from t_0 to $+\infty$, the point (x, y) of a solution moves toward the origin along one of the rays defined by the line $y = 3x$ with the origin deleted.

If the parameter t is eliminated from equation (1.14) or if equation (1.9) is solved, the family of trajectories is given by the equation

$$(x + y)^2 = k(y - 3x).$$

These curves are seen to be tangent to the line $y = 3x$ at the origin (they are parabolas in this particular example, with the lines $2x + 2y + k = 0$ as their axes of symmetry). The equilibrium point at the origin in this case ($q > 0$, $\Delta > 0$) is called a *node*. In this example the node is asymptotically stable.

If the characteristic roots λ_1 and λ_2 had both been positive, the picture would have been similar, but the arrows on the trajectories in Fig. 8.2 would have been reversed, since the movement of a point (x, y) on each trajectory would have been away from the origin as t increased, and the node would be unstable.

Case III. $\Delta < 0$, $p \neq 0$. In this case, the characteristic roots λ_1 and λ_2 are conjugate imaginary numbers, and the equilibrium point at the origin is called a *focus*.

Consider the example

$$\frac{dx}{dt} = x - y,$$
(1.15)
$$\frac{dy}{dt} = x + y.$$

The characteristic roots are $\lambda_1 = 1 + i$, $\lambda_2 = 1 - i$, and the general solution may be written in the form

$$x = e^t(c_1 \cos t + c_2 \sin t)$$
(1.16)
$$y = e^t(c_1 \sin t - c_2 \cos t).$$

The family of curves represented by (1.16) in the phase plane have the equation

(1.17) $$\sqrt{x^2 + y^2} = c e^{\text{arc tan } y/x},$$

where c is an arbitrary constant. The form of this equation suggests the introduction of polar coordinates ρ and θ. Since in polar coordinates,

$$\rho = \sqrt{x^2 + y^2},$$

$$\tan \theta = \frac{y}{x},$$

equation (1.17) becomes

$$\rho = ce^\theta,$$

which is the equation of a spiral.

The situation is described in Fig. 8.3. The motion is away from the equilibrium point because the real part of the characteristic roots is positive, and the focus is unstable. If the real part of the characteristic roots were negative, the arrow would be reversed in Fig. 8.3, and the focus would be asymptotically stable.

When the roots of the characteristic equations are real and unequal, the general case is strictly analogous to the examples given above. When the roots are complex conjugates, the general case requires additional analysis.

Consider the system

(1.18)
$$\dot{x} = ax + by \qquad (b \neq 0),$$

$$\dot{y} = cx + dy.$$

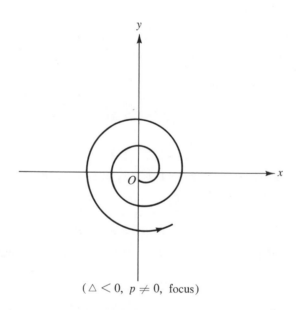

$(\triangle < 0, \ p \neq 0, \ \text{focus})$

FIG. 8.3

Let the characteristic equation

$$\lambda^2 - (a + d)\lambda + (ad - bc) = 0$$

have the roots $\lambda = \alpha \pm i\beta$ $(\beta > 0)$. Then $\alpha^2 + \beta^2 = ad - bc > 0$, and $a + d = 2\alpha$.

We shall employ the following generalization of polar coordinates. Distance from the origin will be measured by

$$\rho = \sqrt{[(\alpha - a)x - by]^2 + \beta^2 x^2},$$

while θ is defined by the equations

(1.19)
$$\beta x = \rho \cos \theta,$$
$$(\alpha - a)x - by = \rho \sin \theta.$$

We have

(1.20)
$$\beta \dot{x} = \dot{\rho} \cos \theta - \rho \dot{\theta} \sin \theta = \beta(ax + by)$$
$$= \rho(\alpha \cos \theta - \beta \sin \theta).$$

Similarly,

$$b\dot{y} = -\dot{\rho} \sin \theta - \rho \dot{\theta} \cos \theta + \frac{\alpha - a}{\beta} \rho(\alpha \cos \theta - \beta \sin \theta) = b(cx + dy)$$

(1.21)
$$= \rho\left(\frac{bc - ad + \alpha d}{\beta} \cos \theta - d \sin \theta\right).$$

From equations (1.20) and (1.21) we have

$$\dot{\rho} \cos \theta - \rho \dot{\theta} \sin \theta = \rho(\alpha \cos \theta - \beta \sin \theta),$$
$$\dot{\rho} \sin \theta + \rho \dot{\theta} \cos \theta = \rho(\beta \cos \theta + \alpha \sin \theta).$$

It follows that

(1.22)
$$\dot{\rho} = \alpha\rho$$
$$\dot{\theta} = \beta \qquad (\beta > 0).$$

Thus

$$\frac{d\rho}{d\theta} = \frac{\alpha}{\beta} \rho,$$

and the trajectories in the phase plane are the spirals

$$\rho = ke^{(\alpha/\beta)\theta}.$$

We have chosen our coordinates so that θ increases with t, and ρ increases or decreases with t according as α is positive or negative, respectively. When $\alpha = 0$, $\dot{\rho} = 0$, and the trajectories are the closed curves $\rho = $ constant; that is, the ellipses

(1.23) $(ax - by)^2 + \beta^2 x^2 = k^2,$

where $k(>0)$ is constant.

In the example

(1.24)
$$\dot{x} = -y,$$
$$\dot{y} = x,$$

$a = 0$, $b = -1$, $c = 1$, $d = 0$, $\lambda = \pm i$, $\alpha = 0$, $\beta = 1$, ρ is the usual polar distance $\sqrt{x^2 + y^2}$, and θ is the usual polar angle. The trajectories (1.23) become, of course, the circles

$$x^2 + y^2 = k^2.$$

This system (1.24) is representative of the interesting case when the characteristic roots are pure imaginaries $\pm i\beta$ ($\beta > 0$). Then $\alpha = (a + d)/2 = 0$, and the trajectories are ellipses having the equation

(1.25) $(ax + by)^2 + (ad - bc)x^2 = k^2 \qquad (d = -a, ad - bc > 0)$

in the phase plane. Typical is the system

$$\dot{x} = 4x + 5y,$$
$$\dot{y} = -5x - 4y.$$

The characteristic equation is

$$\begin{vmatrix} 4 - \lambda & 5 \\ -5 & -4 - \lambda \end{vmatrix} = \lambda^2 + 9 = 0,$$

and the characteristic roots are $\lambda_1 = 3i$, $\lambda_2 = -3i$.

In this case, the general solution may be written in the form

$$x = c_1(5 \cos 3t) + c_2(5 \sin 3t),$$

$$y = c_1(-4 \cos 3t - 3 \sin 3t) + c_2(3 \cos 3t - 4 \sin 3t).$$

The trajectories in the phase plane have the equation

$$5x^2 + 8xy + 5y^2 = c^2,$$

which gives a family of concentric curves (here ellipses) having the origin as center. The equilibrium point at the origin when $\Delta < 0$, $p = 0$, is called a *center* (see Fig. 8.4). Centers are said to be *stable* for systems of the form (1.1), since the trajectories remain bounded.

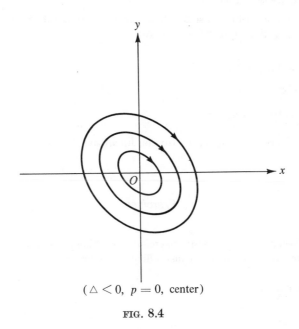

$(\Delta < 0,\ p = 0,\ \text{center})$

FIG. 8.4

When sketching trajectories of systems

$$\dot{x} = ax + by$$
$$\dot{y} = cx + dy,$$

it is frequently useful to observe that along a ray $y = hx$,

$$\frac{dy}{dx} = \frac{c + dh}{a + bh};$$

that is, all trajectories cross a given ray at the same angle, and in the same direction, since

$$\dot{x} = x\,(a + dh)$$
$$\dot{y} = x\,(c + dh).$$

In particular, in sketching a typical trajectory for the system (1.15), one observes that the trajectory that crosses the x-axis at $(1, 0)$ does so with

$\dot{x} = 1$, $\dot{y} = 1$ at that point. Accordingly, all trajectories have slope 1 where they cross the positive x-axis, and motion at those points is upward and to the right. This observation coupled with the fact that $\alpha > 0$ for (1.15) indicates that each trajectory is expanding as it winds about the origin in a counterclockwise direction.

The discussion of the case $\Delta < 0$ is complete. In addition we have analyzed the cases $\Delta > 0$, $q \neq 0$. There remain

1. $(\Delta > 0)$, $q = 0$, $p \neq 0$;

2. $\Delta = 0$, $p \neq 0$;

3. $(\Delta = 0)$, $q = 0$, $p = 0$.

These cases are less important (in the linear case), and we shall not discuss them.

Exercises

Discuss the systems in Exercises 1–6. Sketch typical trajectories and determine the stability or instability of the origin.

1. $\dfrac{dx}{dt} = x + y$,

$\dfrac{dy}{dt} = 3x - y$.

2. $\dfrac{dx}{dt} = x - y$,

$\dfrac{dy}{dt} = 2x + 4y$.

3. $\dfrac{dx}{dt} = -x - y$,

$\dfrac{dy}{dt} = x - y$.

4. $\dfrac{dx}{dt} = 3x + 5y$,

$\dfrac{dy}{dt} = -5x - 3y$.

5. The differential equation

$$\frac{d^2x}{dt^2} + \mu(x^2 - 1)\frac{dx}{dt} + x = 0 \qquad (\mu \text{ a positive constant})$$

is known as van der Pol's equation. It is of importance in vacuum tube theory. The substitution $v = \dfrac{dx}{dt}$ leads to the pair of equations

(1)
$$\frac{dx}{dt} = v,$$
$$\frac{dv}{dt} = -x + \mu(1 - x^2)v.$$

The *equations of variation* of the system (1) are the associated linear equations

(2)
$$\frac{dx}{dt} = v,$$
$$\frac{dv}{dt} = -x + \mu v.$$

Discuss the system (2) and draw some typical trajectories in the xv-plane (the phase plane). Divide your discussion into three cases according as $\mu < 2$, $\mu = 2$, $\mu > 2$.

6. Transform the following pairs of equations by means of (1.19) and obtain equations corresponding to (1.22). Then find equations of the trajectories in "polar" form.

(a) $\dot{x} = 2x + y,$
    $\dot{y} = -x + 2y;$

(b) $\dot{x} = -2x + y,$
    $\dot{y} = -x - 2y;$

(c) $\dot{x} = -2x + y,$
    $\dot{y} = -5x + 2y;$

(d) $\dot{x} = -x - y,$
    $\dot{y} = 10x - 3y;$

(e) $\dot{x} = x + 3y,$
    $\dot{y} = -6x + 7y.$

Answers

1. $\lambda = 2, -2$; saddle point; unstable.
2. $\lambda = 2, 3$; node; unstable.
3. $\lambda = -1 \pm i$; focus; asymptotically stable.
4. $\lambda = 4i, -4i$; center; stable.

2 More general systems

Consider the autonomous system

$$\dot{x} = f(x, y),$$

(2.1)

$$\dot{y} = g(x, y),$$

where $f(x, y)$ and $g(x, y)$ are of class C' in an open domain D containing the origin, and $f(0, 0) = g(0, 0) = 0$. If, at the origin

$$\frac{\partial f}{\partial x} = a, \qquad \frac{\partial f}{\partial y} = b,$$

$$\frac{\partial g}{\partial x} = c, \qquad \frac{\partial g}{\partial y} = d,$$

equations (2.1) can be written in the form

$$\dot{x} = ax + by + F(x, y),$$

(2.2)

$$\dot{y} = cx + dy + G(x, y).$$

The linear equations

$$\dot{x} = ax + by,$$

(2.3)

$$\dot{y} = cx + dy$$

are called the *equations of variation* associated with (2.1) and (2.2). It is natural to ask under what conditions the behavior of the trajectories of (2.3) near the origin is representative of the behavior of the trajectories of (2.2) near the origin. It is of particular importance to ascertain conditions under which the stability of the origin in (2.3) implies stability of the origin for (2.2). We shall examine questions of this kind in the next chapter.

9

Stability; Liapunov functions

1 Nonlinear differential systems

An enormous number of practical problems in the control of electrical, mechanical, thermal, and other engineering systems may be reduced to the study of a system of differential equations of the type

$$\frac{dx_1}{dt} = X_1(x_1, x_2, \ldots, x_n, t),$$

(1.1)
$$\frac{dx_2}{dt} = X_2(x_1, x_2, \ldots, x_n, t),$$

$$\cdot \quad \cdot \quad \cdot \quad \cdot \quad \cdot \quad \cdot \quad \cdot \quad \cdot$$

$$\frac{dx_n}{dt} = X_n(x_1, x_2, \ldots, x_n, t),$$

where the quantities X_i are functions of the $n + 1$ variables $x_1, x_2, \ldots, x_n, t$. The problems that arise in the study of the behavior of solutions of such a system are also of considerable mathematical interest. In this chapter we shall examine some of these questions.

If $x_1, x_2, \ldots, x_n$ are regarded as components of an n-vector x and $X_1, X_2, \ldots, X_n$ as components of an n-vector X, equation (1.1) may be written in abbreviated form as

(1.2)
$$\frac{dx}{dt} = X(x, t).$$

Recall that when the variable t does not appear formally in the right-hand member of (1.1), the system (1.1) is said to be an *autonomous* system; otherwise, it is called *nonautonomous*. When (1.2) is autonomous, it can be rewritten as

(1.2)′
$$\frac{dx}{dt} = X(x).$$

A solution

$$x_1 = x_1(t),$$
$$x_2 = x_2(t),$$
$$\cdot \quad \cdot \quad \cdot$$
$$x_n = x_n(t)$$

of (1.2)′ may be regarded either as a curve in the space of the $n + 1$ variables $x_1, x_2, \ldots, x_n, t$ or as a curve in the space of n variables $x_1, x_2, \ldots, x_n$ with t regarded as a parameter. In the latter case (which will be our principal concern), the curve is called a *trajectory* in the *phase space*—the space of the n variables $x_1, x_2, \ldots, x_n$.

For example, the autonomous system

(1.3)
$$\frac{dx_1}{dt} = x_2,$$
$$\frac{dx_2}{dt} = -x_1$$

has the solution

(1.4)
$$x_1 = \sin t,$$
$$x_2 = \cos t.$$

The solution (1.4) will be recognized as a helix lying on the cylinder $x_1^2 + x_2^2 = 1$ in (x_1, x_2, t)-space, while the corresponding trajectory in the phase space (i.e., the x_1x_2-plane) is the circle $x_1^2 + x_2^2 = 1$. In the latter case t is regarded as a parameter.

Any point $(a_1, a_2, \ldots, a_n)$ that provides a solution of the system of equations

$$X_1(x_1, x_2, \ldots, x_n) = 0,$$
$$\cdot \quad \cdot \quad \cdot \quad \cdot \quad \cdot \quad \cdot$$
$$X_n(x_1, x_2, \ldots, x_n) = 0$$

is called an *equilibrium point* of the system (1.2)′. It is important to study the behavior of trajectories in the neighborhood of an equilibrium point and for this purpose it is usually convenient to assume that the equilibrium point is at the origin. The translation

$$x_i = a_i + z_i \qquad (i = 1, 2, \ldots, n)$$

transforms the system

(1.5) $$\frac{dx}{dt} = X(x)$$

into the equivalent system

(1.6) $$\frac{dz}{dt} = Z(z),$$

where

$$Z(z) = X(a + z),$$

and an equilibrium point $(a_1, a_2, \ldots, a_n)$ of the system (1.5) becomes the equilibrium point $(0, 0, \ldots, 0)$ of the system (1.6).

Exercises

1. Find the solution of the linear system

    $$\frac{dx_1}{dt} = x_2,$$

    $$\frac{dx_2}{dt} = -4x_1,$$

    with the property that $x_1(0) = 0$, $x_2(0) = 1$. Interpret the solution in (x_1, x_2, t)-space and find the corresponding trajectory in the phase plane.

2. Find the solution of the linear system

    $$\frac{dx_1}{dt} = x_1,$$

    $$\frac{dx_2}{dt} = 2x_2$$

    with the property that $x_1(0) = 1$, $x_2(0) = 1$. Interpret the solution in (x_1, x_2, t)-space and find the corresponding trajectory in the phase plane.

3. Show that the equilibrium points of the nonlinear system

    $$\frac{dx_1}{dt} = 8x_1 - x_2^2,$$

    $$\frac{dx_2}{dt} = x_2 - x_1^2$$

    are the points $(0, 0)$ and $(2, 4)$. Transform the system into an equivalent system in which the equilibrium point $(2, 4)$ has been translated to the origin.

4. Show that the equilibrium points of the nonlinear system

$$\frac{dx_1}{dt} = 5 - x_1^2 - x_2^2,$$

$$\frac{dx_2}{dt} = x_2 - 2x_1$$

are the points (1, 2) and $(-1, -2)$. Transform the system to an equivalent system in which the equilibrium point $(-1, -2)$ has been translated to the origin.

5. Find the equilibrium points of the nonlinear system

$$\frac{dx_1}{dt} = x_1 + x_2 - 5,$$

$$\frac{dx_2}{dt} = x_1 x_2 - 6.$$

6. Find the equilibrium points of the nonlinear system

$$\frac{dx_1}{dt} = 293 - x_1^2 - x_2^2,$$

$$\frac{dx_2}{dt} = x_1 x_2 - 34.$$

7. Find the equilibrium points of the nonlinear system

$$\frac{dx_1}{dt} = x_1(x_2 + x_3) - 12,$$

$$\frac{dx_2}{dt} = 6 - x_2(x_1 + x_3),$$

$$\frac{dx_3}{dt} = x_3(x_1 + x_2) - 10.$$

Answers

2. $x_1 = e^t$, $x_2 = e^{2t}$; a curve lying on the surface $x_2 = x_1^2$; the curve $x_2 = x_1^2$.

3. The transformation $x_1 = 2 + z_1$, $x_2 = 4 + z_2$ leads to the equivalent system

$$\frac{dz_1}{dt} = 8z_1 - 8z_2 - z_2^2,$$

$$\frac{dz_2}{dt} = z_2 - 4z_1 - z_1^2.$$

7. $(4, 1, 2)$, $(-4, -1, -2)$.

2 Stability and instability

In the remainder of this chapter we shall deal with autonomous systems of the form

$$\frac{dx_1}{dt} = X_1(x_1, x_2, \ldots, x_n),$$

$$\frac{dx_2}{dt} = X_2(x_1, x_2, \ldots, x_n),$$

(2.1)

$$\cdot \quad \cdot \quad \cdot \quad \cdot \quad \cdot \quad \cdot \quad \cdot$$

$$\frac{dx_n}{dt} = X_n(x_1, x_2, \ldots, x_n),$$

or, in vector form,

(2.1)′
$$\frac{dx}{dt} = X(x).$$

We shall always suppose that there is a unique solution of this system through each point (x^0, t_0) of an appropriate domain D_{n+1} in the $(n + 1)$-space of the variables $x_1, x_2, \ldots, x_n, t$, and, moreover, that this solution exists uniquely for all $t \geq t_0$. It will follow that there is a unique trajectory through each nonequilibrium point of the corresponding domain D of the phase space that is defined for all $t \geq t_0$. We shall assume further that $X(0) = 0$ and that the origin is an *isolated equilibrium point* of the system (2.1)—that is, there exists a sphere with center at the origin such that no other point for which $X(x)$ is zero lies inside the sphere.

Our principal concern will be motions along trajectories in the phase space. Such studies were initiated in Chapter 8 by considering motions along trajectories in the xy-plane for autonomous linear systems such as

$$\frac{dx}{dt} = y,$$

(2.2)

$$\frac{dy}{dt} = -2x - 3y.$$

Here, as there, we shall be particularly concerned with the stability or instability of such motions in a neighborhood of the equilibrium point at the origin.

The equilibrium point at the origin of the system (2.1) is said to be *stable* if corresponding to every positive number R there exists a positive number r ($r \leq R$) such that a motion along a trajectory which, at time $t = t_0$, is at a point inside the sphere $s(r)$

$$x_1^2 + x_2^2 + \cdots + x_n^2 = r^2$$

remains inside the sphere $S(r)$

$$x_1^2 + x_2^2 + \cdots + x_n^2 = R^2$$

for all $t > t_0$. The situation is shown schematically in Fig. 9.1.

The equilibrium point at the origin is said to be *asymptotically stable* in an open domain D containing the origin if it is stable and if every trajectory that starts in D at time $t = t_0$ tends to the origin as $t \to +\infty$. The domain D is then said to be a *domain of attraction* of the equilibrium point at the origin (see Fig. 9.2).

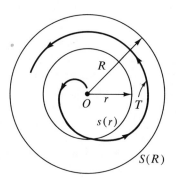

Stable motion

FIG. 9.1

The equilibrium point at the origin is said to be *unstable* when for some R and every positive number r, however small, there is some point P inside $s(r)$ such that the trajectory that is at P at time t_0 will reach the sphere $S(R)$ at time t_1, where t_1 is some number greater than t_0.

When a trajectory is a closed curve, the motion along that trajectory is said to be *periodic*. We observe that the equilibrium point at the origin is stable in Fig. 8.4, unstable in Figs. 8.1 and 8.3, and asymptotically stable in Fig. 8.2. The trajectories in Fig. 8.4 are also examples of periodic motions.

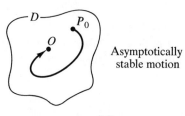

Asymptotically stable motion

FIG. 9.2

3 Characteristic roots

To solve the linear system with constant coefficients

$$\dot{x}_1 = a_{11}x_1 + a_{12}x_2 + \cdots + a_{1n}x_n,$$
$$\dot{x}_2 = a_{21}x_1 + a_{22}x_2 + \cdots + a_{2n}x_n,$$

(3.1)

$$\cdot\quad\cdot\quad\cdot\quad\cdot\quad\cdot\quad\cdot\quad\cdot\quad\cdot\quad\cdot\quad\cdot\quad\cdot$$

$$\dot{x}_n = a_{n1}x_1 + a_{n2}x_2 + \cdots + a_{nn}x_n,$$

recall that we try for a solution of the form

$$x_1 = A_1 e^{\lambda t},$$
$$x_2 = A_2 e^{\lambda t},$$

$$\cdot\quad\cdot\quad\cdot\quad\cdot$$

$$x_n = A_n e^{\lambda t}.$$

This leads to the *characteristic equation*

$$\begin{vmatrix} a_{11} - \lambda & a_{12} & \cdots & a_{1n} \\ a_{21} & a_{22} - \lambda & \cdots & a_{2n} \\ \cdot & \cdot \quad \cdot \quad \cdot \quad \cdot \quad \cdot & \cdot & \cdot \\ a_{n1} & a_{n2} & \cdots & a_{nn} - \lambda \end{vmatrix} = 0.$$

This is a polynomial equation of degree n in λ with real coefficients. The variety of possible roots of such an equation is well known, and the types of terms that appear in the general solution of (3.1) are of the form

(3.2) $$e^{\lambda t} P(t),$$

where λ is a characteristic root (possibly imaginary) and $P(t) \not\equiv 0$ is a polynomial in t. If λ has a positive real part (for example, when $\lambda = 2$, or when $\lambda = 3 + 4i$), some solutions, at least, will not remain bounded as t approaches ∞, and the origin will be unstable. If, however, every characteristic root has a negative real part, every term (3.2) in the solution will tend to 0, as $t \to \infty$, and the origin will be asymptotically stable.

Consider now the nonlinear system

(3.3)
$$\dot{x} = y,$$
$$\dot{y} = -2x - 3y + x^3.$$

The linear terms in (3.3) yield the so-called equations of variation

(3.4)
$$\dot{x} = y,$$
$$\dot{y} = -2x - 3y.$$

The origin is asymptotically stable for the system (3.4), since the characteristic roots $\lambda = -1, -2$ are negative. Inasmuch as the nonlinear term in (3.3) is small near the origin, we would expect that the system (3.4) would determine the stability of the origin in (3.3). It does, and the origin in (3.3) is asymptotically stable. Liapunov theory, introduced in the following section, provides a simple and elegant way of dealing with systems like (3.3). It is effective also in dealing with linear systems having constant coefficients, largely because the task of determining the characteristic roots—or even the signs of their real parts—is often formidable when the characteristic equation is other than quadratic.

Consider, finally, the system

$$(3.5) \qquad \begin{aligned} \dot{x} &= y, \\ \dot{y} &= -3y + kx^3, \end{aligned}$$

where k is a constant $\neq 0$. In this case, the equations of variation are

$$(3.6) \qquad \begin{aligned} \dot{x} &= y, \\ \dot{y} &= -3y, \end{aligned}$$

and while the origin is an isolated equilibrium point for (3.5), it is not for (3.6). In the latter case, the characteristic equation is, of course,

$$\begin{vmatrix} -\lambda & 1 \\ 0 & -3 - \lambda \end{vmatrix} = 0,$$

and the characteristic roots are $\lambda = 0, -3$. The characteristic roots do not give us enough information to determine the stability of the origin for the system (3.5), in which the nonlinearity is the decisive factor. Liapunov theory will tell us rather quickly that when $k < 0$, the origin is asymptotically stable and, almost as quickly, that the origin is unstable when $k > 0$.

4   Liapunov functions

We are ready to formulate criteria that determine whether or not an equilibrium point is stable or unstable. To that end, let

$$V(x) = V(x_1, x_2, \ldots, x_n)$$

be a function of class C' in an open region H containing the origin. Suppose $V(0) = 0$ and that V is positive at all other points of H. Then V has a minimum at the origin, and we say that V is *positive definite* in H. Clearly, the origin is a *critical point* of V—that is, a point at which all the partial derivatives

$$\frac{\partial V}{\partial x_1}, \frac{\partial V}{\partial x_2}, \ldots, \frac{\partial V}{\partial x_n}$$

vanish. The origin will be said to be an *isolated critical point* if there is a sphere with center at the origin such that the origin is the only critical point of V inside the sphere. The *derivative V along trajectories* of (2.1) is defined by the equation

$$\dot{V} = \frac{dV}{dt} = \frac{\partial V}{\partial x_1} X_1(x) + \frac{\partial V}{\partial x_2} X_2(x) + \cdots + \frac{\partial V}{\partial x_n} X_n(x).$$

If $V(x)$ is positive definite in H and if $\dot{V} \leq 0$ throughout H, then $V(x)$ is said to be a *Liapunov function* for the equilibrium point at the origin of the system (2.1)—or, when there is no ambiguity, simply a Liapunov function for the system (2.1).

Example.   Consider the system (2.2) and put

$$V(x, y) = y^2 + 2x^2.$$

Then,

$$\dot{V} = 2y(-2x - 3y) + 4x(y)$$
$$= -6y^2,$$

and the function V is seen to be a Liapunov function for the system (2.2).

Before looking at the geometry in n dimensions let us consider the case $n = 2$. When $V(x, y)$ is positive definite and has the origin as an isolated critical point, it is intuitively evident that the curves

$$V(x, y) = k,$$

for $k > 0$ and sufficiently small, define ovals (closed curves) containing the origin in their interior. Further, when $k_1 < k_2$, the oval defined by $V(x, y) = k_1$ lies entirely within the oval defined by $V(x, y) = k_2$ (see Fig. 9.3). This can be shown to be true, in general, but the proof is not elementary.

We state without proof the *nesting property* of such functions in n-space.

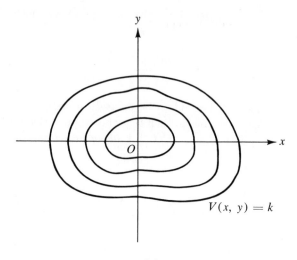

FIG. 9.3

Theorem 4.1 Let $V(x) = V(x_1, x_2, \ldots, x_n)$ be a function of class C' in an (open) region H containing the origin that has the following properties:

(a) $V(0) = 0$,
(b) $V(x) > 0$ for $x \neq 0$,
(c) *the origin is the only critical point of $V(x)$ in H.*

Then, for $k > 0$ sufficiently small, the equation $V(x) = k$ defines a closed surface containing the origin in its interior. Further, if k_1 $(0 < k_1 < k)$ is a constant, the closed surface defined by $V(x) = k_1$ contains the origin in its interior and is wholly contained in the closed surface defined by $V(x) = k$.

If H is all of n-space, and if $V(x)$ has at most a finite number of critical points (finite or infinite), the equation $V(x) = k$ defines a closed surface containing the origin in its interior inside of which the nesting property is valid for each $k < k_0$, where k_0 is the smallest positive value that $V(x)$ assumes at one of its critical points. The equation $V(x) = k_0$ may or may not define a closed surface.

(See, however, Exercise 4 of this section where, it will be observed, $V(x, y)$ has an uncountable infinity of finite critical points, and the conclusion of Theorem 4.1 remains valid.)

Example.   Consider the function

$$V(x, y) = 3x^2 + y^2 - x^3.$$

It is positive definite neighboring the origin, and its only critical points are at $(0, 0)$ and $(2, 0)$. Thus, the critical point at the origin is isolated. Here the number $k_0 = 4$. Consider the curve

(4.1) $$3x^2 + y^2 - x^3 = 2.$$

Its graph is given in Fig. 9.4.

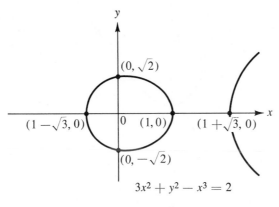

FIG. 9.4

Note that the total curve (4.1) is not an oval, but the equation defines an oval near the origin. The equation

$$3x^2 + y^2 - x^3 = 1,$$

for example, will then define an oval lying wholly within the oval defined by (4.1) and containing the origin in its interior.

In this example, the equation $V(x) = 4$, that is,

$$3x^2 + y^2 - x^3 = 4$$

defines a closed curve surrounding the origin. Its graph appears in Fig. 9.5.

Example. Consider the function

$$V(x, y) = \frac{4x^2}{1 + x^2} + y^2.$$

This function is positive definite in every neighborhood of the origin. Here,

$$V_x = \frac{8x}{(1 + x^2)^2}, \qquad V_y = 2y.$$

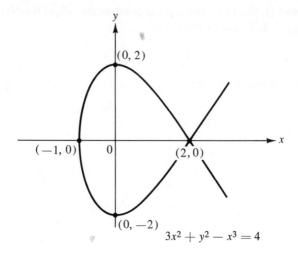

$$3x^2 + y^2 - x^3 = 4$$

FIG. 9.5

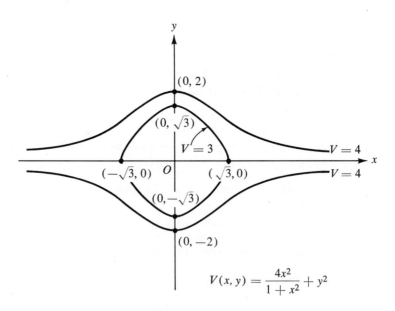

$$V(x, y) = \frac{4x^2}{1 + x^2} + y^2$$

FIG. 9.6

The *critical values* of this function are then

$$V(0, 0) = 0, \qquad V(\infty, 0) = 4, \qquad V(-\infty, 0) = 4.$$

Accordingly, $k_0 = 4$, and the equation $V(x, y) = k \ (0 < k < 4)$ defines a closed curve containing the origin in its interior. The equation $V(x, y) = 4$, that is,

$$\frac{4x^2}{1 + x^2} + y^2 = 4,$$

does not define a closed curve.

Exercises

In each of the following three exercises plot the four curves whose equations are given using the same coordinate axes. In each case the function of x, y involved is positive definite near the origin and has the origin as an isolated minimum point.

1. $x^2 + 2y^2 = k$ $(k = 1, 2, 4, 6)$.

2. $3x^2 + y^2 - x^3 = k$ $(k = 1, 2, 4, 6)$.

3. $\dfrac{3x^2}{1 + x^2} + y^2 = k$ $(k = 1, 2, 3, 4)$.

4. Note that the function

$$V(x, y) = 5(x^2 + y^2) - (x^2 + y^2)^2$$

is positive definite in the region $D:x^2 + y^2 < 5$ and that its level curves $V(x, y) = c$ in D are circles. Discuss the nesting property of the level curves in D.

We are prepared to prove the following fundamental result.

Theorem 4.2.   (LIAPUNOV) If corresponding to the system (2.1), which we write as

(4.2) $\dot{x} = X(x),$

there exists a Liapunov function $V(x)$, the equilibrium point at the origin of the system (4.2) is stable.

To prove the theorem (see Fig. 9.7), let S be a sphere, with center at the origin, which with its interior lies wholly in an (open) domain H in which $V(x)$ is positive definite. Because S is compact (a bounded, closed set), $V(x)$ attains its absolute minimum $m > 0$ on S at some point P_1 on S. Because $V(0) = 0$ and $V(x)$ is continuous inside S, we can choose a ball B bounded by a smaller sphere s, with center at the origin, throughout which $V(x) \leq \delta$, where $0 < \delta < m$. Let P_0 be any point, except the origin, in B, and consider the trajectory T that is at P_0 when $t = t_0$. Note that $0 < V(P_0) \leq \delta$. Because $\dot{V} \leq 0$ along T, V never increases with t; accordingly, $V \leq V(P_0) \leq \delta$, for $t \geq t_0$. Thus, V evaluated along T remains less than m for all $t > t_0$. It follows that T can never reach S, and the theorem is proved.

Note that we have not employed the nesting property of $V(x)$ in the proof of Theorem 4.1, nor have we required that the origin be an isolated critical point of the function $V(x)$.

Liapunov functions may be regarded as extensions of the idea of energy functions in mechanics. But to apply Liapunov's idea, any function $V(x, y)$ satisfying the hypotheses of the theorem will suffice. In simpler cases we may often guess usable Liapunov functions; in more complicated cases determining a Liapunov function for a given system may be very difficult indeed.

As an example consider the system

$$\frac{dx}{dt} = -y,$$

$$\frac{dy}{dt} = x.$$

The point $(0, 0)$ is the only equilibrium point of the system. A suitable Liapunov function is $V(x, y) = x^2 + y^2$. For then

$$\dot{V} = 2x(y) + 2y(-x) = 0.$$

Thus, the critical point at the origin is stable. The trajectories are, of course, the family of circles $x^2 + y^2 = a^2$.

Theorem 4.3.   (LIAPUNOV) *If, in addition to the hypotheses of Theorem 4.2, the function* $-\dot{V} \in C'$ *is positive definite† and both V and* $-\dot{V}$ *ve the origin as an isolated critical point, the equilibrium point at the origin is asymptotically stable.*

Note first that the origin is stable.

Next, suppose that we are concerned with trajectories that for $t \geq t_0$ lie in a neighborhood N of the origin in which the nesting property is valid for both $V(x)$ and $-\dot{V}(x)$ (see Fig. 9.8). Let $k_1 > 0$ be chosen small enough that the closed surface S determined by the equation $V(x) = k_1$ lies in N. Let T be a trajectory that at $t = t_0$ is at a point P_0, not the origin, inside S. Because V is strictly decreasing along T, as t increases, V tends to a limit k $(0 \leq k < k_1)$, as $t \to \infty$. It will be sufficient to show that $k = 0$.

---

† A function $W(x)$ is said to be negative definite when $-W(x)$ is positive definite.

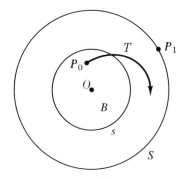

FIG. 9.7

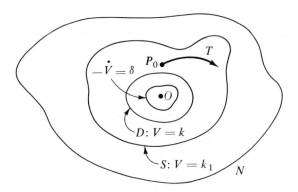

FIG. 9.8

Suppose, on the contrary, that $k > 0$. Then T cannot reach the closed surface D determined by the equation $V(x) = k$, because $V(x) > k$, for $t \geq t_0$. Finally, let $\delta > 0$ be taken small enough that the closed surface s determined by the equation $-\dot{V} = \delta$ lies wholly inside D. Recall that $-\dot{V} < \delta$ inside s and that $-\dot{V} > \delta$ outside s. Then, $\dot{V} < -\delta$ on T for $t \geq t_0$. It follows that

$$(4.3) \qquad \int_{t_0}^{t} \dot{V}\, dt < \int_{t_0}^{t} -\delta\, dt \qquad (t > t_0),$$

or†

$$V(t) < V(t_0) - \delta(t - t_0).$$

---

† When it is convenient to do so and it can cause no real difficulty, we shall use $V(t)$, $V(x)$, and $V(P)$ interchangeably. The slight ambiguity is more than compensated for by the resulting simplicity of notation.

This would imply that $V(t) \to -\infty$, as $t \to \infty$, which is impossible. Accordingly, $k = 0$, and the theorem is proved.

Asymptotic stability, when it occurs, is a local phenomenon that is associated with a neighborhood of an equilibrium point. The neighborhood may, to be sure, be very large, possibly the entire phase space.

As an example of the application of Theorem 4.3, consider the system (see Fig. 8.2)

$$\frac{dx}{dt} = -\frac{7}{4}x + \frac{1}{4}y,$$

$$\frac{dy}{dt} = \frac{3}{4}x - \frac{5}{4}y.$$

The origin is the only equilibrium point. A suitable Liapunov function is

$$V(x, y) = 12x^2 + 12xy + 20y^2.$$

The equations $V(x, y) = k$ $(k > 0)$ represent a family of ellipses. We compute

$$\dot{V} = -(33x^2 + 47y^2);$$

accordingly, Theorem 4.3 yields the fact that the equilibrium point at the origin is asymptotically stable.

It will be observed that an equally good choice (among others) for V would have been $3x^2 + 6xy + 11y^2$.

We continue with the following theorem on instability.

Theorem 4.4. (LIAPUNOV) *Let $V(x)$ be of class C'' in a region H containing the origin and suppose that $V(0) = 0$. If $\dot{V}$ (evaluated along trajectories) is positive definite in H and has the origin as an isolated critical point, and if in every neighborhood of the origin there is a point where $V > 0$, the equilibrium point at the origin is unstable.*

Before a proof of this theorem is given, a few comments on the hypotheses may be helpful. Although the positive-definiteness of a function plays a basic role in classical Liapunov methods, in general it is the nesting property of positive-definite functions that is crucial, since positive-definiteness alone is not sufficient to ensure that the function has the nesting property. The critical point at the origin must be isolated.

To prove the theorem we proceed as follows. Let $k > 0$ be such that the equation $\dot{V} = k$ defines a closed surface $H_1 \subset H$ within which the nesting property for the function $\dot{V}$ holds. Next, let S and s be any two spheres in H_1 with centers at the origin and such that s lies inside S. Pick a point P_0 inside

s with the property that $V(P_0) > 0$ and consider the trajectory T commencing at P_0 at time t_0. Since $\dot{V} > 0$ inside and on S, except at the origin, V increases steadily along T, as t increases. Because of the continuity of V, there is a closed spherical domain B, centered at the origin, throughout which $|V(x)| \leq V(P_0)/2$. Note that T cannot reach any point of B.

Next, take $\delta > 0$ so small that the closed surface determined by $\dot{V} = \delta$ lies wholly within B. Then, on T,

$$\dot{V} > \delta \qquad (t \geq t_0),$$

and, hence,

$$\int_{t_0}^{t} \dot{V}\, dt > \int_{t_0}^{t} \delta\, dt;$$

that is, along T,

$$V(t) - V(t_0) > \delta(t - t_0) \qquad (t \geq t_0).$$

Because of the continuity of V in the closed ball bounded by S, V is bounded there, but by the last inequality, V cannot remain bounded. Accordingly, T must pass outside S, and the origin is unstable (see Fig. 9.9).

To illustrate the theorem consider the system (see Fig. 8.1)

$$(4.4) \qquad \frac{dx}{dt} = x - y, \qquad \frac{dy}{dt} = -2x.$$

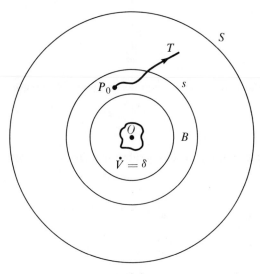

FIG. 9.9

The function $V(x, y) = x^2 - 2xy - y^2$ satisfies the hypotheses of the theorem, for at every point on the line $2x + y = 0$, except the origin, we note that $V > 0$, and that along each trajectory

$$\dot{V} = 2x\dot{x} - 2x\dot{y} - 2y\dot{x} - 2y\dot{y}$$
$$= 6x^2 + 2y^2.$$

Accordingly, the equilibrium point at the origin is unstable.

In practice, contrary to the traditional definition of a Liapunov function given above, it is becoming customary nowadays so to designate any function V used, as above, to determine the stability or instability of an equilibrium point.

This method of studying the stability or instability of an equilibrium point is called *Liapunov's second method* or *Liapunov's direct method*.

Exercises

1. Find a Liapunov function for each system in Exercises 1–4 of Chapter 8, Section 1, and in each case determine the stability or instability of the equilibrium point of the system. [*Hint.* Assume $V = x^2 + kxy + my^2$, where k and m are constants to be determined. Recall that a form $ax^2 + bxy + cy^2$ is definite if and only if $b^2 - 4ac < 0$.]

2. Discuss the stability of the equilibrium point at the origin of the system

$$\frac{dx}{dt} = -y,$$

$$\frac{dy}{dt} = 3x.$$

3. Discuss the stability of the equilibrium point at the origin of the system

$$\dot{x} = -2x + y,$$
$$\dot{y} = 3x + 4y.$$

4. Discuss the stability of the equilibrium point at the origin of the nonlinear system

$$\frac{dx}{dt} = y + x^3 + xy^2,$$

$$\frac{dy}{dt} = -x + x^2y + y^3.$$

Take $V = x^2 + y^2$.

5. Discuss the stability of the equilibrium point at the origin for the system

$$\dot{x} = -2x - y,$$
$$\dot{y} = 10x + 4y.$$

5 Stability of linear systems with constant coefficients

Consider the system

(5.1)
$$\dot{x} = y + x^2 - x^3 y,$$
$$\dot{y} = -2x - 3y - xy + x^3.$$

The origin is an isolated equilibrium point of (5.1), and the corresponding equations of variation are

(5.2)
$$\dot{x} = y,$$
$$\dot{y} = -2x - 3y.$$

The characteristic roots of (5.2) are $\lambda = -1, -2$. The origin is then an asymptotically stable equilibrium point of (5.2). Near the origin, the non-linear terms $x^2 - x^3 y, -xy + x^3$ in (5.1) are small relative to the linear terms, and we anticipate that the origin is also asymptotically stable in (5.1). And this is true. Indeed, we may employ the same Liapunov function for both systems. Such a function is

(5.3)
$$2v = 5x^2 + 2xy + y^2.$$

Then, for (5.2),

$$\dot{V} = -2(x^2 + y^2);$$

accordingly, the origin in (5.2) is asymptotically stable.

Applying the same function $2V$ to (5.1), we have

$$\dot{V} = -2(x^2 + y^2) + [xy^2 + 5x^3 + x^3 y + x^4 - x^3 y^2 - 5x^4 y].$$

Inasmuch as the terms in the bracket are of third degree and higher near the origin, the function $-2(x^2 + y^2)$ is dominant. It follows that the origin, in (5.1), is asymptotically stable.

It is natural then to consider the problem of determining a Liapunov function for a system

(5.4)
$$\dot{x} = Ax,$$

where all the characteristic roots of the matrix $A = (a_{ij})$ have negative real parts. In the analysis that follows we shall employ a minimal amount of linear algebra. In particular, if H' denotes the transpose of a matrix H, recall that

$$(GH)' = H'G'.$$

Also, if $G(t)$ and $H(t)$ are differentiable matrices, then

$$(GH)^{\cdot} = \dot{G}H + G\dot{H}.$$

Suppose then that B is an $n \times n$ symmetric matrix $B = (b_{ij})$, where $b_{ij} = b_{ji}$ (that is, $B' = B$). Then

$$V = x'Bx$$

is a symmetric quadratic form, and its derivative along trajectories of (5.4) is given by

$$
\begin{aligned}
\dot{V} &= (\dot{x})'Bx + x'B\dot{x} \\
&= x'A'Bx + x'BAx \\
&= x'(A'B + BA)x.
\end{aligned}
$$

Thus $\dot{V}$ is a quadratic form with matrix $A'B + BA = -C$.

We shall show next that C is symmetric—that is, $C' = C$. We have at once that

$$C' = -(A'B + BA)' = -(BA + A'B) = C.$$

Accordingly, if $V = x'Bx$ is any symmetric quadratic form, its derivative along trajectories of (5.4) is given by the quadratic form $-x'Cx$, where C is also a symmetric matrix.

Suppose now that all the characteristic roots of the matrix A in (5.4) have negative real parts and that the matrix C is the unit matrix; that is, $C = (\delta_{ij})$, where $\delta_{ij} = 0$ $(i \neq j)$, and $\delta_{ii} = 1$. We shall show that there then exists a solution B of the matrix equation

(5.5) $$A'B + BA = -C(= -I),$$

where the matrix B is symmetric, and the quadratic form

$$Q = x'Bx$$

is positive definite. Thus Q will be a Liapunov function for the system (5.4) that will establish the origin as an asymptotically stable equilibrium point.

To that end consider the system

(5.6) $$\dot{y} = -yA$$

adjoint to (5.4). Here y is a row vector, and the characteristic roots of this system are the negatives of those of the system (5.4); accordingly, all characteristic roots of (5.6) have positive real parts. Further, there exists a unique set of n linearly independent solutions of (5.6) that can be written as the rows of a matrix $Y(t)$, where

$$Y(0) = I.$$

Recall that then the determinant

(5.7) $$|Y(t)| \neq 0,$$

for all t, because the n solutions in $Y(t)$ are linearly independent.

Each element of a particular solution in $Y(t)$ is of the form $e^{\mu t} f(t)$, where $\mu > 0$, and $f(t)$ may be a polynomial in t or such a polynomial plus a finite number of terms of the type $t^m \sin ht$ and $t^m \cos ht$, where m is a nonnegative integer and h is a constant. Thus the matrix

$$B = \int_{-\infty}^{0} Y'(t) C Y(t)\, dt$$

exists,† and we shall show that it is the solution of (5.5) that we seek.

First, observe that

$$[Y'(t) C Y(t)]' = Y'(t) C Y(t)$$

for all t; thus B is symmetric. Next, we have

$$A'B = \int_{-\infty}^{0} A'Y'CY\, dt = \int_{-\infty}^{0} (YA)'CY\, dt = -\int_{-\infty}^{0} \dot{Y}'CY\, dt,$$

$$BA = \int_{-\infty}^{0} Y'CYA\, dt = -\int_{-\infty}^{0} Y'C\dot{Y}\, dt,$$

and

$$A'B + BA = -\int_{-\infty}^{0} (\dot{Y}'CY + Y'C\dot{Y})\, dt = -\int_{-\infty}^{0} (Y'CY)'\, dt$$

$$= -[Y'(t)CY(t)]_{-\infty}^{0} = -C.$$

---

† Because the analysis that follows goes through substantially unchanged when C is the matrix of any symmetric, positive-definite quadratic form, we defer replacing C by I as long as possible.

It follows that B is a solution of (5.5).

It remains to show that the quadratic form $x'Bx$ is positive definite. To see this, observe that the quadratic form

$$(5.8) \qquad\qquad\qquad y'Cy$$

is positive definite; that is, it is positive for all nonnull vectors y. Set

$$(5.9) \qquad\qquad\qquad y = Y(t)x.$$

Because of (5.7) the vector x is zero, if and only if y is the null vector. Further,

$$y' = (Y(t)x)' = x'Y'(t),$$

and if the substitution (5.9) is made in (5.8), we see that the quadratic form

$$x'Y'(t)CY(t)x$$

is, for each t, positive unless x is the null vector. Finally, then, the quadratic form

$$x'\left[\int_{-\infty}^{0} Y'(t)CY(t)\,dt \right]x = x'Bx$$

is positive definite.

The following result, due to Sylvester, is useful in the applications. A proof can be found in most textbooks on linear algebra.

Theorem 5.1. (SYLVESTER) *If B is a symmetric matrix, a necessary and sufficient condition that the quadratic form $x'Bx$ be positive definite is that the n principal determinants of B be positive.*

The *principal determinants are the set*

$$|b_{11}|, \begin{vmatrix} b_{11} & b_{12} \\ b_{21} & b_{22} \end{vmatrix}, \begin{vmatrix} b_{11} & b_{12} & b_{13} \\ b_{21} & b_{22} & b_{23} \\ b_{31} & b_{32} & b_{33} \end{vmatrix}, \ldots, \begin{vmatrix} b_{11} & \cdots & b_{1n} \\ b_{21} & \cdots & b_{2n} \\ \cdot & \cdot\cdot\cdot & \cdot \\ b_{n1} & \cdots & b_{nn} \end{vmatrix}.$$

Consider next the system

$$\dot{x}_1 = a_{11}x_1 + a_{12}x_2 + \cdots + a_{1n}x_n + f_1(x_1, x_2, \ldots, x_n),$$

$$\dot{x}_2 = a_{21}x_1 + a_{22}x_2 + \cdots + a_{2n}x_n + f_2(x_1, x_2, \ldots, x_n),$$

(5.10)

$$\cdots \cdots \cdots \cdots \cdots \cdots \cdots \cdots \cdots$$

$$\dot{x}_n = a_{n1}x_1 + a_{n2}x_2 + \cdots + a_{nn}x_n + f_n(x_1, x_2, \ldots, x_n),$$

where the a_{ij} are real constants and the functions f_i are of class C' in some spherical ball B with center at the origin with $f_i(0) = 0$ $(i = 1, 2, \ldots, n)$. We suppose further that the origin is an isolated equilibrium point of the system (5.10). Equations (5.10) may be written in vector form as

(5.11) $$\dot{x} = Ax + f(x).$$

Theorem 5.2. If the eigenvalues of $A(x)$ have negative real parts and if, in addition to the above hypotheses on $f(x)$, there exists a constant M such that $\|f(x)\| \leq M\|x\|^2$, the origin is an asymptotically stable equilibrium point of the system (5.11).

Recall that $\|f\| = \sqrt{f_1^2 + f_2^2 + \cdots + f_n^2}$, the euclidean norm† of f.

To prove the theorem, we require the following two lemmas.

Lemma. Let $B = (b_{ij})$ be an $n \times n$ matrix and let $\bar{b}$ be the maximum of the set of n^2 absolute values $|b_{ij}|$. Then,

(5.12) $$\|Bx\| \leq \bar{b}\sqrt{n} \sum_{i=1}^{n} |x_i|.$$

To prove the lemma note that

$$\|Bx\| = [(b_{11}x_1 + \cdots + b_{1n}x_n)^2 + (b_{21}x_1 + \cdots + b_{2n}x_n)^2 + \cdots$$
$$+ (b_{n1}x_1 + \cdots + b_{nn}x_n)^2]^{1/2}$$

and that each of the n terms in the parentheses is less than

$$\bar{b}^2(|x_1| + |x_2| + \cdots + |x_n|)^2.$$

Thus

$$\|Bx\| \leq [n\bar{b}^2(|x_1| + |x_2| + \cdots + |x_n|)^2]^{1/2},$$

and (5.12) follows.

---

† For a very useful elementary treatment of matrix and vector norms, see H. W. Brinkmann and E. A. Klotz, *Linear Algebra and Analytic Geometry*, Chap. 10, Addison-Wesley, Reading, Mass. (1971).

A similar argument establishes the inequality

$$(5.13) \qquad \|x'B\| \le \bar b \sqrt{n} \sum_{i=1}^{n} |x_i|.$$

Lemma. (SCHWARZ INEQUALITY) *If y is a row vector and z is a column vector, both having n components, then*

$$(5.14) \qquad \|yz\| \le \|y\| \cdot \|z\|$$

The proof of this fundamental result is very easy. If a and b are any real numbers,

$$\sum_{i=1}^{n} (ay_i + bz_i)^2 \ge 0;$$

that is,

$$\sum_{i=1}^{n} (a^2 y_i^2 + 2aby_i z_i + b^2 z_i^2) = a^2 \sum_{i=1}^{n} y_i^2 + 2ab \sum_{i=1}^{n} y_i z_i + b^2 \sum_{i=1}^{n} z_i^2 \ge 0.$$

Because the last inequality holds for all a and b, the discriminant

$$\left[\sum_{i=1}^{n} y_i z_i \right]^2 - \sum_{i=1}^{n} y_i^2 \sum_{i=1}^{n} z_i^2 \le 0,$$

and (5.14) follows.

We return to the proof of Theorem 5.2. Let B be the solution of the matrix equation

$$A'B + BA = -I.$$

Recall that because the characteristic roots of A have negative real parts, B is symmetric and the quadratic form

$$(5.15) \qquad V = x'Bx$$

is positive definite. We evaluate $\dot V$ along trajectories of (5.11) and have, successively,

$$
\begin{aligned}
\dot V &= \dot x' Bx + x' B \dot x \\
&= [Ax + f]'Bx + x'B[Ax + f] \\
(5.16) \qquad &= [x'A' + f']Bx + x'B[Ax + f] \\
&= x'(A'B + BA)x + [f'Bx + x'Bf] \\
&= -\|x\|^2 + [f'(Bx) + (x'B)f].
\end{aligned}
$$

Applying the Schwarz inequality to the terms in the bracket above, we have, in turn,

$$\|f'(bx) + (x'B)f\| \leq \|f'(Bx)\| + \|(x'B)f\|$$

$$\leq [\|f'\| + \|f\|]\bar{b}\sqrt{n} \sum_{i=1}^{n} |x_i|$$

$$\leq 2\|f\|\bar{b}\sqrt{n} \sum_{i=1}^{n} |x_i|,$$

since $\|f'\| = \|f\|$.

Finally, since $\|f\| \leq M\|x\|^2$, it follows that the norm of the bracketed terms in (5.16) is less than

$$2M\|x\|^2\bar{b}\sqrt{n} \sum_{i=1}^{n} |x_i|.$$

Thus

$$\dot{V} \leq -\|x\|^2 \left[1 - 2M\bar{b}\sqrt{n} \sum_{i=1}^{n} |x_i|\right];$$

accordingly, $-\dot{V}$ is positive definite inside the domain containing the origin that is determined by the inequality

$$\sum_{i=1}^{n} |x_i| < \frac{1}{2M\bar{b}\sqrt{n}}.$$

The proof of the theorem is complete.

We state without proof the following result, which is complementary to Theorem 5.2.

Theorem 5.3. If $f(x)$ has the properties given in Theorem 5.2 and if at least one eigenvalue of A has a positive real part, the origin in (5.11) is unstable.

In practice, when solving the matrix equation $A'B + BA = C$ for B, where A and C are given matrices, one ordinarily employs algebraic methods. For example, to find B, when

$$A = \begin{bmatrix} 0 & 1 \\ -2 & -3 \end{bmatrix}, \quad C = \begin{bmatrix} -2 & 0 \\ 0 & -2 \end{bmatrix},$$

let

$$B = \begin{bmatrix} b_{11} & b_{12} \\ b_{12} & b_{22} \end{bmatrix},$$

where the numbers b_{ij} are to be determined. We have, successively, that

$$\begin{bmatrix} 0 & -2 \\ 1 & -3 \end{bmatrix}\begin{bmatrix} b_{11} & b_{12} \\ b_{12} & b_{22} \end{bmatrix} + \begin{bmatrix} b_{11} & b_{12} \\ b_{12} & b_{22} \end{bmatrix}\begin{bmatrix} 0 & 1 \\ -2 & -3 \end{bmatrix} = \begin{bmatrix} -2 & 0 \\ 0 & -2 \end{bmatrix},$$

$$\begin{bmatrix} -2b_{12} & -2b_{22} \\ b_{11} - 3b_{12} & b_{12} - 3b_{22} \end{bmatrix} + \begin{bmatrix} -2b_{12} & b_{11} - 3b_{12} \\ -2b_{22} & b_{12} - 3b_{22} \end{bmatrix} = \begin{bmatrix} -2 & 0 \\ 0 & -2 \end{bmatrix},$$

$$\begin{bmatrix} -4b_{12} & b_{11} - 3b_{12} - 2b_{22} \\ b_{11} - 3b_{12} - 2b_{22} & 2b_{12} - 6b_{22} \end{bmatrix} = \begin{bmatrix} -2 & 0 \\ 0 & -2 \end{bmatrix}.$$

From the last equation we obtain the following equations for the determination of the numbers b_{ij}:

$$-4b_{12} \qquad\qquad = -2,$$
$$b_{11} - 3b_{12} - 2b_{22} = \quad 0,$$
$$2b_{12} - 6b_{22} = -2.$$

A little algebra then yields the fact that

$$B = \begin{bmatrix} \frac{5}{2} & \frac{1}{2} \\ \frac{1}{2} & \frac{1}{2} \end{bmatrix}.$$

Exercises

1.   Solve the matrix equation $A'B + BA = -I$ for B, when

(a) $A = \begin{bmatrix} 0 & 1 \\ -1 & -2 \end{bmatrix};$

(b) $A = \begin{bmatrix} -7 & 1 \\ 3 & -5 \end{bmatrix};$

(c) $A = \begin{bmatrix} -1 & -1 \\ 1 & -1 \end{bmatrix}.$

2.   Solve the matrix equation $A'B + BA = -2I$ for B, when

(a) $A = \begin{bmatrix} 0 & 2 \\ -3 & -5 \end{bmatrix};$

(b) $A = \begin{bmatrix} -1 & 1 \\ 0 & -2 \end{bmatrix};$

(c) $A = \begin{bmatrix} -1 & 1 \\ -2 & 1 \end{bmatrix}.$

3. Solve the matrix equation $A'B + BA = -2I$ for B, when

$$A = \begin{bmatrix} 0 & 1 & 0 \\ 0 & 0 & 1 \\ -1 & -1 & -2 \end{bmatrix}.$$

It will probably simplify matters somewhat to take B in the form

$$B = \begin{bmatrix} a & b & c \\ b & d & e \\ c & e & f \end{bmatrix}.$$

4. See the comment in Exercise 3 and solve the matrix equation

$$A'B + BA = -2I$$

for B, when

$$A = \begin{bmatrix} 1 & 1 & 0 \\ 0 & 0 & 1 \\ -5 & -4 & -2 \end{bmatrix}.$$

5. Show that if $A = \begin{bmatrix} a_{11} & a_{12} \\ a_{21} & a_{22} \end{bmatrix}$ and B is taken in the form $B = \begin{bmatrix} a & b \\ b & c \end{bmatrix}$, the

solution of the matrix equation $A'B + BA = -2I$ leads to the equations

$$a_{11}a + a_{21}b = -1,$$
$$a_{12}a + (a_{11} + a_{22})b + a_{21}c = 0,$$
$$a_{12}b + a_{22}c = -1,$$

for the determination of a, b, and c. Show that these equations can always be solved for a, b, and c provided that the negative of a characteristic root of A is not also a characteristic root of A.

6. For which of the following matrices A can the matrix equation $A'B + BA = -I$ be solved for B? If any can be solved, does the resulting quadratic form $x'Bx$ lead to a determination of the stability or instability of the origin of the system $\dot{x} = Ax$?

$$\begin{bmatrix} 1 & -1 \\ 2 & 4 \end{bmatrix}, \quad \begin{bmatrix} 1 & 1 \\ 3 & -1 \end{bmatrix}, \quad \begin{bmatrix} 3 & 5 \\ -5 & 3 \end{bmatrix}, \quad \begin{bmatrix} 1 & -1 \\ -1 & -1 \end{bmatrix}, \quad \begin{bmatrix} 1 & 1 \\ 2 & 2 \end{bmatrix}.$$

7. Determine the stability or instability of the equilibrium point at the origin of the system

$$\dot{x} = y,$$
$$\dot{y} = z - xy,$$
$$\dot{z} = -x - y - 2z + 2x^2.$$

(*Hint.* Use Exercise 3.)

8.   Determine the stability or instability of the equilibrium point at the origin of the system

$$\dot{x} = x + y - x^2y + 2yz,$$
$$\dot{y} = z + x^2,$$
$$\dot{z} = -5x - 4y - 2z + z^3.$$

9.   Use any available method to determine the stability or instability of the origin of the system

$$\dot{x} = y - x^2,$$
$$\dot{y} = z + 2xy - z^2,$$
$$\dot{z} = 4x - 3z - y^3.$$

Answers

3. $a = 8, b = 6, c = 1, d = 17, e = 7, f = 4.$

4. $a = 64, b = 43, c = 13, d = 33, e = 11, f = 6.$

7.   Choose $M = 5$ in Theorem 5.2, for example.

6   The theorems of LaSalle and of Četaev

The theorems of Liapunov in Section 4 are of considerable theoretical and historical interest. In this section we shall develop a theorem on asymptotic stability due to LaSalle and a theorem on instability due to Četaev that are more recent and that are basic.

We begin with LaSalle's theorem. It is concerned with a system (2.1)

(6.1) $\dot{x} = X(x)$

having an isolated equilibrium point at the origin.

Before stating LaSalle's theorem, we require certain preliminaries (we shall follow LaSalle's treatment†).

---

† J. P. LaSalle, "Some extensions of Liapunov's second method," *IRE Trans. Prof. Group on Circuit Theory*, Vol. CT-7 (1960), pp. 520–527.

Let $x(t)$ be a solution of (6.1). A solution will be represented geometrically by the trajectory T_0 in the phase space determined by the equations

$$x_i = x_i(t),$$

where t is regarded as a parameter. Following Birkhoff‡ we shall say that a point p is in the *ω-limit set* Γ^+ of $x(t)$, if corresponding to each number $\varepsilon > 0$ and each $T > 0$ there is a number $t > T$ with the property that $\|x(t) - p\| < \varepsilon$. That is to say, that for each such point p there exists a sequence t_n ($n = 1, 2, 3, \ldots$) tending to infinity such that $x(t_n) \to p$, as $n \to \infty$.

It can be shown§ that if $x(t)$ is bounded for $t \geq 0$, Γ^+ is a nonempty, compact, connected, invariant set. A set M is said to be (positive) *invariant* if each solution starting in M at $t = t_0$ remains in M for all $t > t_0$.

Next, a trajectory $x(t)$ is said to approach a set M, as $t \to \infty$, if for each $\varepsilon > 0$ there exists a number $T > 0$ such that for all $t > T$, each point of $x(t)$ is within a distance ε of some point of M.

Lemma 6.1. If $x(t)$ is bounded for $t > t_0$, $x(t)$ approaches its ω-limit set Γ^+, as $t \to \infty$.

Suppose the contrary. There would then exist an $\varepsilon > 0$ such that for each $T > 0$ there would be a number $t > T$ such that $\|x(t) - p\| \geq \varepsilon$, for all p in Γ^+. Thus, there would be a sequence $t_n \to \infty$, as $n \to \infty$, such that $\|x(t_n) - p\| \geq \varepsilon$, for all p in Γ^+. But $x(t)$ is bounded; accordingly, the sequence $x(t_n)$ has a limit point in Γ^+—which is a contradiction.

It follows from Lemma 6.1 that if $x(t)$ is bounded for $t \geq 0$, and if a set M contains Γ^+, then $x(t) \to M$, as $t \to \infty$.

We are now prepared to prove the following theorem.

Theorem 6.1. (LASALLE) *Let $\bar{H}$ be a bounded closed set with the property that every trajectory of (6.1) that begins in $\bar{H}$ remains in $\bar{H}$, as $t \to \infty$. Suppose that there exists a function $V(x)$ of class C' in $\bar{H}$ for which $\dot{V}(x) \leq 0$ in $\bar{H}$. Let E be the set of points in $\bar{H}$ where $\dot{V}(x) = 0$. If M is the largest invariant set in E, every trajectory starting in $\bar{H}$ tends to M, as $t \to \infty$.*

Let T_0 be a trajectory that starts in $\bar{H}$ at $t = t_0$. Since $\dot{V}(x) \leq 0$, V never increases along T_0, as t increases. The function $V(x)$ is bounded below in $\bar{H}$. Thus V tends to a limit c, along T_0, as $t \to \infty$. Observe that

‡ G. D. Birkhoff, "Dynamical systems," *Amer. Math. Soc. Colloq. Publ.*, Vol. 9 (1927), pp. 197–198.
§ Ibid.

$\Gamma^+(T_0) \subset \bar{H}$, since $\bar{H}$ is closed and that, because of the continuity of V in $\bar{H}$, $V(x) \equiv c$ on Γ^+. It follows that $\dot{V}(x) \equiv 0$ on Γ^+ and, therefore, that $\Gamma^+ \subset M$. Accordingly, $T_0 \to M$, as $t \to \infty$.

The proof of the theorem is complete.

The application of LaSalle's theorem usually lies along the following lines. One seeks a function $V(x)$ such that for some positive number k, the set $V(x) \leq k$ is a bounded set $\bar{H}$. If $\dot{V}(x) \leq 0$ in $\bar{H}$, $V(x)$ never increases along any solution that starts in $\bar{H}$. Then, if $\bar{H}$ has the nesting property, a trajectory that starts in $\bar{H}$ must remain in $\bar{H}$.

We have the following result.

Corollary. Let $\bar{H}$ be a bounded closed region defined by $V(x) \leq k$ and suppose that $V(x)$ is of class C' in $\bar{H}$. If $\dot{V}(x) \leq 0$ in $\bar{H}$ and if $V(x)$ has the nesting property, every solution starting in $\bar{H}$ approaches M, as $t \to \infty$.

Let us illustrate the use of the corollary. Consider the system

(6.2)
$$\dot{x} = y,$$
$$\dot{y} = -2x - y + x^2$$

and take

$$V = 2x^2 + y^2 - \tfrac{2}{3}x^2.$$

We note that the origin is an isolated equilibrium point of the system (6.2) and that V is positive definite near the origin and has the origin as an isolated critical point. Further, $V(x)$ is bounded in the domain bounded by the closed curve defined by the equation

(6.3)
$$2x^2 + y^2 - \tfrac{2}{3}x^3 = \tfrac{8}{3}.$$

We compute

$$\dot{V} = -2y^2.$$

Thus $\dot{V} \leq 0$, and $\dot{V} = 0$ only along the x-axis. By definition, an invariant set contained in the set where $\dot{V} = 0$—that is, in the line $y = 0$—must be such that a trajectory starting at a point of the x-axis will remain in the x-axis. A reference to the system (6.2) reveals that no part of the line $y = 0$ is a solution, except the origin. Accordingly, the origin is the set M of Theorem 6.1. It follows that the origin is asymptotically stable.

A companion theorem to Theorem 6.1 is due to Četaev. It is this.

Theorem 6.2. (ČETAEV) If there exists a bounded, closed neighborhood N of the origin, an open region $N_1 \subset N$, and a function $V(x)$ such that

(a) $V(x)$ is of class C' in N, with $V(0) = 0$;
(b) the origin lies on the boundary ∂N_1 of N_1, but not on $\partial N_1 \cap \partial N$;
(c) $V(x) > 0$ for $x \in N_1$, $V(x) = 0$ for $x \in \partial N_1 - \partial N$;
(d) $\dot{V}(x) \geq 0$ for $x \in N_1$;
(e) the set of points $\{[N_1 + (\partial N_1 \cap \partial N)]| \dot{V}(x) = 0\}$ contains no positive invariant set of (6.1) (see Fig. 9.10),

the equilibrium point at the origin in (6.1) is unstable.

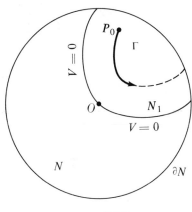

FIG. 9.10

In part c the symbol $\partial N_1 - \partial N$ means the boundary of N_1 with boundary points of N deleted.

To prove the theorem consider a trajectory T starting at $P_0 \in N_1$ at time $t = t_0$. Note that $V(P_0) > 0$. Since $\dot{V} \geq 0$ in N_1, $V(T) \geq V(P_0)$ as long as the trajectory remains in N_1. Further, for all $t > t_0$, T is bounded away from that part of the boundary of N_1 that lies wholly within N, inasmuch as $V = 0$ there. It is sufficient to show that T must reach the boundary ∂N of N in finite time. If it did not, by the argument used in the proof of LaSalle's theorem, the set $\{[N_1 + (\partial N_1 \cap \partial N)]| \dot{V}(x) = 0\}$ would contain the invariant set $\Gamma^+(T)$ on which $\dot{V}(x) = 0$, contrary to (e).

The proof is complete.

Example. Consider the isolated equilibrium point at the origin of the system.

(6.4)
$$\dot{x} = y$$
$$\dot{y} = x - y,$$

and the function $V = x^2 - y^2$. Let N be bounded by any circle with center at the origin. The open set N_1 may be taken as the set bounded by the

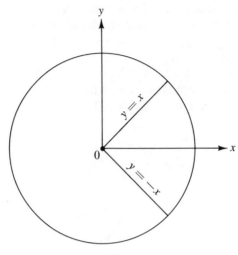

FIG. 9.11

circle and the rays $y = x$ and $y = -x (x \geq 0)$. The derivative

$$\dot{V} = y^2$$

is positive, except along the x-axis where it is zero. The only invariant set in the set of points for which $\dot{V} = 0$ is the origin. Accordingly, the hypotheses of the theorem are satisfied, and the origin is, therefore, an unstable equilibrium point of the system (6.4). Note that an alternative form of the theorem may be obtained by changing $V(x) > 0$ and $\dot{V}(x) \geq 0$ to $V(x) < 0$ and $\dot{V}(x) \leq 0$, respectively, in its statement. As will be noted in the example, this is simply equivalent to replacing $V(x)$ by $-V(x)$. We state this observation as a corollary.

Corollary. If the conditions $V(x) > 0$ and $\dot{V}(x) \geq 0$ in Četaev's theorem are replaced, respectively, by $V(x) < 0$ and $\dot{V}(x) \leq 0$, the conclusion of Četaev's theorem remains valid.

The problem of determining the stability, or instability, of an equilibrium point at the origin of our system

$$\frac{dx}{dt} = X(x)$$

leads then to a search for a suitable function $V(x)$ that will enable us to apply either LaSalle's theorem, or Četaev's. And this search, in general, is very difficult. For systems

$$\dot{x} = y,$$
$$\dot{y} = f(x, y) \qquad [f(0, 0) = 0]$$

a very useful formula is available.† It is the following:

(6.5) $$V(x, y) = y^2 - 2 \int_0^x f(x, 0) \, dx.$$

In many important cases this leads to a function that enables one to apply either Theorem 6.1 or 6.2.

Example. Consider the differential equation

$$\ddot{x} + g(x) = 0,$$

where $g(x)$ is of class C' for all x, and $g(0) = 0$. If we set $\dot{x} = y$, this differential equation will be replaced by the equivalent system

$$\dot{x} = y,$$
$$\dot{y} = -g(x).$$

If we now employ (6.5), we have

$$V(x, y) = y^2 + 2 \int_0^x g(x) \, dx$$

and

$$\dot{V} = 2y[-g(x)] + 2g(x)y \equiv 0.$$

If $xg(x) > 0$ $(x \neq 0)$, V is positive definite throughout the xy-plane. Because $\dot{V} \equiv 0$, the origin is stable. Further, since $\dot{V} \equiv 0$, V is a constant

---

† "On the construction of Liapunov functions for certain autonomous nonlinear differential equations," *Contrib. Differential Equations*, Vol. 2 (1963), pp. 367–383. Formula (6.5) has been generalized for systems

$$\dot{x} = f(x, y)$$
$$\dot{y} = g(x, y),$$

when the equation $f(x, y) = 0$ has the solution $y = h(x)$. It becomes then

$$V = \int_{h(x)}^y f(x, y) \, dx - \int_0^x g[x, h(x)] \, dx.$$

See L. R. Anderson and W. Leighton, "Liapunov functions for autonomous systems of second order," *J. Math. Analysis & Appl.*, Vol. 23 (1968), pp. 645–664.

along trajectories; that is, along trajectories we have

$$y^2 + 2 \int_0^x g(x)\, dx = k^2 \qquad (k \text{ constant}).$$

This last is then, in fact, an equation of the trajectories of the system. The trajectories are accordingly periodic motions, at least when k^2 is sufficiently small.

Domains of attraction.   Consider the system

(6.6)
$$\dot{x} = y,$$
$$\dot{y} = -2x - x^2 - y + y^2.$$

This system has equilibrium points at $(0, 0)$ and at $(-2, 0)$. To study the stability of the origin, let

$$V(x, y) = y^2 + 2 \int_0^x (2x + x^2)\, dx$$

$$= y^2 + 2x^2 + \frac{2x^3}{3}.$$

This function is positive definite near the origin. The equation

$$V(x, y) = V(-2, 0)$$

or

(6.7)
$$y^2 + 2x^2 + \frac{2x^3}{3} = \frac{8}{3}$$

defines a closed curve about the origin, and by Theorem 4.1, $V(x, y)$ possesses the nesting property inside (6.7). Along trajectories

$$\dot{V} = 2y(-2x - x^2 - y + y^2) + 4xy + 2x^2y$$
$$= -2y^2(1 - y).$$

The only trajectory included in the points for which $\dot{V} = 0$ is the origin, and $\dot{V} \leq 0$ for $y \leq 1$.

LaSalle's theorem almost applies. The problem is that the line $y = 1$ cuts through the closed region defined by (6.7). This difficulty may be resolved by choosing $k > 0$ so that the curve

$$y^2 + 2x^2 + \frac{2x^3}{3} = k$$

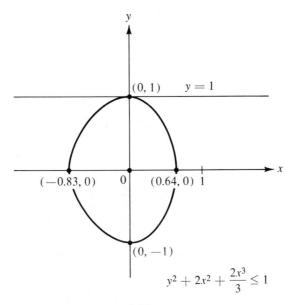

$$y^2 + 2x^2 + \frac{2x^3}{3} \leq 1$$

FIG. 9.12

is tangent to the line $y = 1$. The choice $k = 1$ is immediate, and it follows that LaSalle's theorem applies to the interior of the closed curve defined by the equation

$$(6.8) \qquad\qquad y^2 + 2x^2 + \frac{2x^3}{3} = 1.$$

Every trajectory that commences inside the oval defined by (6.8) remains inside that oval. The origin is asymptotically stable, and the interior of the oval defined by equation (6.8) is a domain of attraction for the equilibrium point at the origin of the system (6.6) (see Fig. 9.12).

We return to equations (6.6) to examine the equilibrium point at $(-2, 0)$. It will be convenient to transform the differential equations by means of the equations

$$x = u - 2, \qquad y = y$$

that translate the point $(-2, 0)$ to the origin. Equations (6.6) become

$$(6.6)' \qquad \begin{aligned} \dot{u} &= y, \\ \dot{y} &= 2u - y - u^2 + y^2. \end{aligned}$$

We again employ the formula (6.5):

$$V(u, y) = y^2 - 2 \int_0^u (2u - u^2) \, du$$

$$= y^2 - 2u^2 + \frac{2u^3}{3}.$$

Then,

$$\dot{V} = 2y(2u - y - u^2 + y^2) - 4uy + 2u^2y$$
$$= -2y^2(1 - y).$$

The curve $V = 0$ divides the circular domain $u^2 + y^2 \leq a^2$ $(a^2 < 1)$ into four quadrants (see Fig. 9.13). In the open quadrant N_1 containing the positive x-axis, $V < 0$ and $\dot{V} \leq 0$. The Corollary to Četaev's theorem applies, and the point $(-2, 0)$ is an unstable equilibrium point of the system (6.6).

It is clear that one could replace $V(u, y)$ above by $-V(u, y)$ and apply Četaev's theorem.

To study the behavior of solutions of the nonlinear second-order equation

$$\ddot{x} + \dot{x} - \dot{x}^2 + 2x + x^2 = 0,$$

for example, it is appropriate to set $\dot{x} = y$, and the equivalent pair of equations (6.6) results.

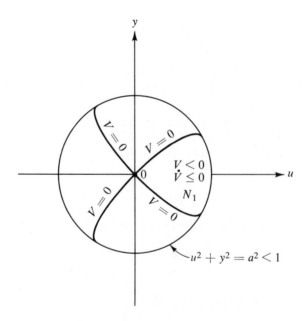

FIG. 9.13

The principles we have employed in the case of two first-order autonomous differential equations are valid for systems of n such equations, when one can determine a suitable function $V(x)$.

Recapitulation. LaSalle's theorem is clearly of prime importance for the determination of domains of attraction of an isolated asymptotically stable equilibrium point of autonomous systems

$$(6.9) \qquad\qquad \dot{x} = X(x).$$

We make the usual assumptions on $X(x)$ including the assumption that the equilibrium point in question is at the origin.

To apply the theorem, one must determine a function $V(x)$ of class C' neighboring the origin such that for some constant $k > 0$ the equation $V(x) = k$ determines a bounded closed set $\bar{H}$ that contains a ball with center at the origin. Further, the nesting property valid for $V(x)$ in $\bar{H}$ will ensure that trajectories that start in $\bar{H}$ remain in $\bar{H}$. Next, k must be chosen sufficiently small that $\dot{V}(x) \leq 0$ inside $\bar{H}$.

We conclude this section with a statement of the following theorem, the proof of which is omitted.

Theorem 6.3. Let $V(x)$ be a positive-definite function in the whole of n-space that has its only critical point at the origin. If $\dot{V}(x) \leq 0$ throughout n-space with the origin as the only trajectory along with $\dot{V} = 0$, and if $V(x) \to \infty$, as $\|x\| \to \infty$, every solution is bounded, the origin is asymptotically stable, and its domain of attraction is all of n-space.

Exercises

1. Discuss the stability of the equilibrium point of the system

$$\frac{dx}{dt} = y,$$

$$\frac{dy}{dt} = -2x + ky^3 \qquad (k \text{ constant}).$$

2. The substitution $\dot{x} = y$ in van der Pol's equation

$$\ddot{x} + \mu(x^2 - 1)\dot{x} + x = 0$$

leads to the nonlinear system

$$\frac{dx}{dt} = y,$$

$$\frac{dy}{dt} = -\mu(x^2 - 1)y - x.$$

Using $V = x^2 + y^2$, show that the equilibrium point at the origin is asymptotically stable when $\mu < 0$. Find a domain of attraction of the origin.

3.   Discuss the stability of the equilibrium points of the system

$$\dot{x} = y,$$
$$\dot{y} = -6x - y - 3x^2.$$

If either point is asymptotically stable, find a domain of attraction for it. In addition, if such a domain is given by $V(x, y) = k_0$, sketch the three curves $V(x, y) = k_0, k_0 + 1, k_0 - 1$.

4.   Discuss the stability of each equilibrium point of the system

$$\dot{x} = y,$$
$$\dot{y} = -4x - y - 4x^2.$$

If either equilibrium point is asymptotically stable, find a domain of attraction.

5.   Discuss the stability of the equilibrium points of the system

$$\frac{dx}{dt} = y,$$

$$\frac{dy}{dt} = -x - x^2 - y^3.$$

If either equilibrium point is asymptotically stable, find a domain of attraction.

6.   If $ad - bc \neq 0$, the origin is an isolated equilibrium point of the system

$$\frac{dx}{dt} = ax + by \qquad (b \neq 0),$$

$$\frac{dy}{dt} = cx + dy.$$

Use the function $V(x, y) = (ax + by)^2 + (ad - bc)x^2$ to show that if $a + d < 0$ and $ad - bc > 0$, the origin is asymptotically stable. Show also that these inequalities are equivalent to the condition that the real parts of the roots of the characteristic equation

$$\begin{vmatrix} a - \lambda & b \\ c & d - \lambda \end{vmatrix} = 0$$

be negative.

7.   Discuss the stability of each equilibrium point of the system

$$\dot{x} = y,$$
$$\dot{y} = 4x - y - x^2 + 2y^2.$$

If either is asymptotically stable, find a domain of attraction.

8.   Show that the origin is an asymptotically stable equilibrium point of the differential system

$$\dot{x} = -2x + y,$$
$$\dot{y} = -2x + y - x^3.$$

9.   Show that the origin is a stable equilibrium point of the differential system

$$\dot{x} = -x + y,$$
$$\dot{y} = -2x + y - x^3.$$

10.   Study the equilibrium points of the system

$$\dot{x} = y,$$
$$\dot{y} = -3x - x^2 - y + y^2.$$

If the system has an asymptotically stable equilibrium point, find a domain of attraction for it.

11.   Consider the system

$$\dot{x} = y,$$
$$\dot{y} = z,$$
$$\dot{z} = -2y - 3z - 2x^3,$$

which has the origin as an isolated equilibrium point. Compute the characteristic roots of the associated equations of variation. Regroup terms to show that near the origin the function

$$2V = 3x^4 + 11y^2 + z^2 + 4x^3y + 6yz$$

is positive definite and that along trajectories

$$\dot{V} = -6y^2(1 - x^2).$$

Finally, use LaSalle's theorem to establish the asymptotic stability of the origin and find a domain of attraction of the origin.

12. Study the stability or instability of the equilibrium point at the origin for the system

$$\dot{x} = y,$$
$$\dot{y} = z,$$
$$\dot{z} = -2y - 2z - x^3$$

by means of the function

$$2V = z^2 + 4yz + 6y^2 + 2x^3y + x^4.$$

If the origin is asymptotically stable, find a domain of attraction.

13. Do the same as in Exercise 12 for the system

$$\dot{x} = y,$$
$$\dot{y} = z,$$
$$\dot{z} = -2y - 2z - kx^3 \qquad (k \text{ constant}),$$

using the function

$$2V = z^2 + 4yz + 6y^2 + 2kx^3y + kx^4.$$

14. Consider the differential equation

$$\ddot{x} + a\ddot{x} + b\dot{x} + f(x) = 0,$$

where a and b are positive constants, $f(x)$ is of class C' on an appropriate interval I with $xf(x) > 0$ for $x \neq 0$, and $f(0) = 0$. Transform this equation to a system of three differential equations by means of the substitution $\dot{x} = y$, $\dot{y} = z$, obtaining the system

$$\dot{x} = y,$$
$$\dot{y} = z,$$
$$\dot{z} = -by - az - f(x),$$

and study the equilibrium point at the origin of the resulting system by means of the function

$$2V = b(ay + z)^2 + [by + f(x)]^2 + 2 \int_0^x f(x)[ab - f'(x)] \, dx.$$

Show that if $f'(x) < ab$ for $|x|$ small, the origin is asymptotically stable. (In particular, it follows that if $f(x) = cx$ $(c > 0)$, a sufficient condition that the origin be asymptotically stable is $ab > c$, a so-called *Routh-Hurwitz condition*.†) Finally, find a domain of attraction of the origin, when $f'(x) < ab$ on I, and $f(x)$ vanishes only at $x = 0$ and at $x = x_0 > 0$.

15.   Show that the origin is an asymptotically stable equilibrium point of the system

$$\dot{x} = y,$$

$$\dot{y} = z,$$

$$\dot{z} = -2x - 2y - 3z - 2x^2,$$

and find a domain of attraction.

16.   Use the function $2V$ in Exercise 14 to study the stability of the equilibrium point at the origin of the system

$$\dot{x} = y,$$

$$\dot{y} = z,$$

$$\dot{z} = -3y - 2z - x^2.$$

17.   Set $\dot{x} = y$, $\dot{y} = z$, and $\dot{z} = w$ in the differential equation

$$\ddddot{x} + 6\dddot{x} + 11\ddot{x} + 6\dot{x} + 4x^3 = 0,$$

and show that the origin of the resulting system is asymptotically stable.

Answer

17.   One possible Liapunov function is

$$2V = w^2 + 38z^2 + 135y^2 + 12wz + 120zy + 18wy + 48x^3y + 8x^3z + 18x^4.$$

---

† See, for example, J. V. Uspensky, *Theory of Equations*, pp. 304–309, McGraw-Hill, New York (1948).

10

Quadratic functionals

In this chapter we shall study the important connections between self-adjoint linear differential equations of second order and their associated quadratic functionals. Although this material is properly part of the calculus of variations, the methods we shall employ will be elementary.

1 The fundamental problem

The value of the integral

$$(1.1) \qquad J(Y) = \int_0^1 (y'^2 - y^2)\, dx$$

is determined by the particular function $y(x)$ for which it is evaluated. Thus, when $y = 1$, $J = -1$, and when $y = x$, $J = \frac{2}{3}$. Because the value of J depends on the particular function $y(x)$ employed, J is an example of a *functional*. The integrand is a quadratic in the symbols y and y'; because of this, (1.1) is also an example of *quadratic functional*.

The quadratic functional which we shall consider is

$$(1.2) \qquad J(y) = \int_a^b [r(x)y'^2 - p(x)y^2]\, dx,$$

where $r(x) > 0$, and $r(x)$ and $p(x)$ are continuous on the (closed) interval $[a, b]$. We shall be concerned with the value of J when $y(x)$ is any function

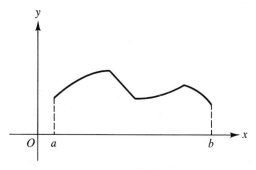

FIG. 10.1

of class D' on $[a, b]$ such that $y(a) = y(b) = 0$. Such a function $y(x)$ will be
said to be *admissible*. A function is of class D' when it is continuous on $[a, b]$
and it is of class C' on this interval, except possibly for a finite number of
corners at each of which it has finite left- and right-hand derivatives. The
graph of such a function (with three corners) is given in Fig. 10.1

Note that a function of class C' on $[a, b]$ is also of class D' there.

The fundamental problem. Is there an admissible function $y(x)$ that
affords a minimum value to J? If so, what is the function, and what is the
corresponding value of J?

The last question is readily answered. We observe that if any admissible
function yields a minimum to J, the minimum value of J must be zero.
For the function 0 is admissible, and $J(0) = 0$. Consequently, the minimum
value of J cannot be positive. The minimum value of J also cannot be
negative. For suppose $J(y_1) = k < 0$, and that $y_1(x)$ yields a minimum to
J. Then $J(cy_1) = c^2k \ (c > 1)$ would be less than $J(y_1)$.

The two questions above will be considered in the next section.

2   The conjugate point

With the functional

$$(2.1) \qquad J(y) = \int_a^b [r(x)y'^2 - p(x)y^2] \, dx$$

we associate its *Euler-Jacobi equation*

$$(2.2) \qquad [r(x)y']' + p(x)y = 0.$$

We recall that $r(x)$, $p(x)$ are assumed to be continuous with $r(x) > 0$ on

$[a, b]$. The differential equation (2.2) is seen to be in self-adjoint form.

It will be convenient to define $r(x)$ and $p(x)$ on a slightly longer interval $(a - h, b + h)$ $(h > 0)$ by means of the equations

(2.3)
$$
\begin{aligned}
r(x) &= r(a) & (a - h < x < a), \\
p(x) &= p(a) & (a - h < x < a), \\
r(x) &= r(b) & (b < x < b + h), \\
p(x) &= p(b) & (b < x < b + h).
\end{aligned}
$$

Then $r(x) > 0$ and $r(x)$, $p(x)$ are continuous on the longer interval.

Let $y(x)$ be any nonnull solution of (2.2) that vanishes at $x = a$. If there is a number c $(a < c < b + \varepsilon)$ such that $y(c) = 0$ with $y(x) \neq 0$ on (a, c), the point $x = c$ on the x-axis is called the *first conjugate point* of $x = a$, or simply, the conjugate point of $x = a$.

We come to the principal result in this chapter.

Theorem 2.1. If there is no point conjugate to $x = a$ on $(a, b]$, $J > 0$ unless $y = 0$. If there is a point conjugate to $x = a$ on the interval (a, b), J has no minimum. Finally, if $x = b$ is the first conjugate point of $x = a$, $J > 0$ except for solutions $z(x)$ of (2.2) that vanish at $x = a$ and $x = b$. Along these solutions $J = 0$.

To prove the theorem, suppose first that there is no point conjugate to $x = a$ on $(a, b]$. There then exists a solution $z(x)$ of the Euler-Jacobi equation such that

$$
z(b + k) = 0, \qquad z(x) > 0 \qquad (a \leq x \leq b)
$$

for $0 < k < h$, and k sufficiently small. For, let $z_1(x)$ be any solution that vanishes at $x = a$ and is positive on $(a, b]$. Then $z_1(x) > 0$ on $[b, b + 2k]$ for $k > 0$ and sufficiently small. The solution $z(x)$ defined by the conditions

(2.4)
$$
z(b + k) = 0, \qquad z'(b + k) = -1
$$

is the solution we seek (by the Sturm separation theorem).

Next, let $y(x)$ be any admissible function and let $z(x)$ be an arbitrary solution of the Euler-Jacobi equation. On any interval on which $z(x) \neq 0$, it is easy to verify that

(2.5)
$$
ry'^2 - py^2 \equiv r\left(y' - \frac{z'}{z} y\right)^2 + \left(r \frac{z'}{z} y^2\right)'.
$$

If $z(x)$ is the solution defined by (2.4), we have

$$(2.6) \quad J(y) = \int_a^b (ry'^2 - py^2)\, dx = \left[r\frac{z'}{z} y^2 \right]_a^b + \int_a^b r\left(y' - \frac{z'}{z} y \right)^2 dx$$

$$= \int_a^b r\left(y' - \frac{z'}{z} y \right)^2 dx.$$

Thus, $J(y) = 0$, if and only if

$$(2.7) \qquad\qquad\qquad y'(x) - \frac{z'(x)}{z(x)} y(x) \equiv 0$$

between corners of $y(x)$; otherwise $J(y) > 0$. If $y(x)$ has no corners, (2.7) implies that

$$y(x) \equiv cz(x) \qquad (a \le x \le b),$$

where c is a constant. But $y(a) = 0$ and $z(a) \ne 0$. It follows that $c = 0$. If $y(x)$ is assumed to have corners—it can have at most a finite number—the same argument shows that $y(x) \equiv 0$ from $x = a$ to the first corner. Then a repetition of the argument shows that $y(x) \equiv 0$ from the first corner either to the second corner, if there is such, or to $x = b$. It follows, repeating the argument if necessary, that $y(x) \equiv 0$ on $[a, b]$ (and incidentally that it can have no corners).

Accordingly, we have shown that when there is no point on $(a, b]$ conjugate to $x = a$, $J = 0$ only when $y(x) \equiv 0$, and $J > 0$ for all other admissible functions.

Next, we shall suppose there is a point $x = c$ on (a, b) conjugate to $x = a$ and we shall show that there exists an admissible function $y(x)$ for which $J < 0$—and, hence, J can have no minimum.

To that end, let $y(x)$ be a solution that vanishes at $x = a$ and at $x = c$ with $y(x) > 0$ on (a, c). Let $x = d$ be a point near† $x = c$ ($c < d < b$), and consider the solution $z(x)$ defined by the conditions (see Fig. 10.2)

$$z(d) = 0, \qquad z'(d) = -1.$$

Such a solution $z(x)$ is linearly independent of $y(x)$ and must vanish on (a, c); accordingly the curves $y = y(x)$ and $y = z(x)$ must intersect at a point $[h, y(h)]$, where $a < h < c$, and $y(h) = z(h) > 0$.

Because $y(x)$ and $z(x)$ are linearly independent,

$$\Delta(x) = r(x)[y(x)z'(x) - z(x)y'(x)] \equiv c \ne 0.$$

---

† It is conceivable that $x = a$ could have a second conjugate point on (a, b). By "near" is meant any point $x = d$ such that the only point conjugate to $x = a$ on $(a, d]$ is the point $x = c$.

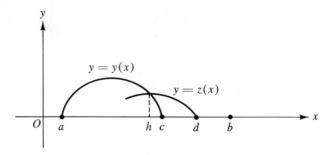

FIG. 10.2

Note that $r(a) > 0$, $y(a) = 0$, $z(a) < 0$, $y'(a) > 0$; thus, $c > 0$. Then, $\Delta(h) > 0$, and it follows that

(2.8) $$z'(h) > y'(h).$$

We shall evaluate J for the admissible function defined by the equations

(2.9) $$y = \begin{cases} y(x) & (a \leq x \leq h), \\ z(x) & (h \leq x < d), \\ 0 & (d \leq x \leq b). \end{cases}$$

An integration by parts yields the fact that

(2.10) $$J(y) = \int_{\alpha}^{\beta} (ry'^2 - py^2)\, dx = [ryy']_{\alpha}^{\beta} - \int_{\alpha}^{\beta} y[(ry')' + py]\, dx,$$

where α and $\beta \in [a, b]$. If $y(x)$ is a solution of the Euler-Jacobi equation on the interval (α, β), equation (2.10) becomes

(2.11) $$J(y) = [ryy']_{\alpha}^{\beta}.$$

Formula (2.11) makes evaluation of J along the curve (2.9) easy. We have

$$J = [ryy']_a^h + [rzz']_h^d + 0$$
$$= r(h)y(h)y'(h) - r(h)z(h)z'(h)$$
$$= r(h)y(h)[y'(h) - z'(h)].$$

The bracketed quantity in the last line is negative, by (2.8), and we have accordingly shown that *a necessary condition that J possess a minimum is that the interval (a, b) contains no point conjugate to $x = a$.*

The proof of Theorem 2.1 will be complete when we have considered the case when $x = b$ is the first conjugate point of $x = a$. For this, we return to equation (2.6) where $y(x)$ is an arbitrary admissible function, and now $z(x)$ is a nonnull solution of the Euler-Jacobi equation that vanishes at $x = a$ and at $x = b$. The ratio y/z has a finite limit as $x \to a$ and as $x \to b$; consequently, we have again that

$$(2.12) \qquad J(y) = \int_a^b r\left(y' - \frac{z'}{z}y\right)^2 dx.$$

It follows that $J > 0$ unless

$$y' - \frac{z'}{z}y \equiv 0;$$

that is, unless

$$y(x) \equiv cz(x),$$

where c is a constant.† Thus, $J = 0$ only along solutions $cz(x)$ of the Euler-Jacobi equation. Conversely, when $y(x) = cz(x)$, it follows at once from (2.10) that $J = 0$.

The proof of the theorem is complete.

Our primary concern is with the differential equation (2.2) and its associated functional (2.1); however, Theorem 2.1 is valid as stated when the functional is

$$(2.13) \qquad J(y) = \int_a^b [a(x)y'^2 + 2b(x)yy' + c(x)y^2]\, dx,$$

where $a(x) > 0$, and $a(x)$, $b(x)$, $c(x)$ are continuous on $[a, b]$. The associated Euler-Jacobi differential equation is then the equation

$$(2.14) \qquad [a(x)y' + b(x)y]' - [b(x)y' + c(x)y] = 0.$$

3 Some examples and applications

Consider first the functional

$$(3.1) \qquad J = \int_0^b (y'^2 - y^2)\, dx.$$

---

† By the type of argument employed above. Of course, in the present situation, it cannot be argued that $c = 0$, inasmuch as $y(a) = z(a) = 0$, but one shows that there *is* a single constant c such that $y(x) \equiv cz(x)$ on $[a, b]$.

The corresponding Euler-Jacobi equation is

(3.2) $$y'' + y = 0.$$

A solution defining the conjugate points of $x = 0$ is $\sin x$, and, consequently, the first conjugate point of $x = 0$ is $x = \pi$. Theorem 2.1 then yields the following results (when J is evaluated in the class of admissible functions):
  1. *If $b > \pi$, J has no minimum;*
  2. *If $b < \pi$, $J > 0$ except along the null solution;*
  3. *If $b = \pi$, $J > 0$ for all admissible functions except the set $c \sin x$ (c constant), for which $J = 0$.*
  Consider the differential equation

(3.3) $$y'' + (7 - x^2)y = 0 .$$

We shall determine the upper bound for the first conjugate point of $x = 0$. To that end, we write the corresponding functional

(3.4) $$J = \int_0^b [y'^2 - (7 - x^2)y^2] \, dx.$$

The function $x(b - x)$ is not a solution of (3.3). It is an admissible function, however. We evaluate J for this function and determine a value b such that $J = 0$. This leads to the equation

(3.5) $$\int_0^b [(b - 2x)^2 - (7 - x^2)(bx - x^2)^2] \, dx = 0,$$

or

$$2b^4 - 49b^2 + 70 = 0.$$

From this we have $b^2 = 1.52$, $b^4 = 2.30$, approximately. The smallest positive value of b that satisfies equation (3.5) is then 1.23, approximately, and this is an upper bound of the first conjugate point of $x = 0$ (the actual conjugate point is known to be $c = \sqrt{1.5}$, so the bound is a very good one).
  The argument goes as follows. The value of b obtained from (3.5) cannot be the actual conjugate point $x = c$, for $x(b - x)$ is not a solution of the differential equation. If c were greater than b, the function

$$y = \begin{cases} bx - x^2 & (0 \le x \le b) \\ 0 & (b < x \le c) \end{cases}$$

would be admissible, $\neq 0$, and would yield the value 0 to J, contrary to the theorem. It follows that $c < b$.

A bound determined by this process is always an *upper* bound for the conjugate point, by a similar argument.

Choices of functions other than $x(b - x)$ are, of course, possible. For example, we might have employed

$$\sin \frac{\pi x}{b}, \qquad x(x^2 - b^2), \qquad x^m(b - x)^n \qquad (m, n \text{ positive integers})$$

or even

$$e^{-x^2/2}(x^3 - b^2 x).$$

The last choice would have been especially fortuitous for one would have discovered that this function is a solution of the differential equation for $b^2 = \frac{3}{2}$, and accordingly yields the exact value of the conjugate point.

Frequently, writing the functional in the form (2.10) leads to somewhat simpler computations. The equation to be solved for b then becomes

(3.6) $$\int_a^b y[(ry')' + py] \, dx = 0,$$

inasmuch as the term outside the integral is zero.

The problem of finding a lower bound for the conjugate point requires different techniques and is, in general, more difficult.

Exercises

1. Employ the function $x(x^2 - b^2)$ to obtain an upper bound for the first conjugate point of $x = 0$ for equation (3.3). Use (3.6).

2. Find an upper bound for the first conjugate point of $x = 0$ for the differential equation

$$y'' + xy = 0.$$

3. Do the same as in Exercise 2 for the differential equation

$$y'' + x^2 y = 0.$$

4. Find an upper bound for the first conjugate point of $x = 1$, when the differential equation is

$$x^2 y'' + xy' + y = 0.$$

5. Find an upper bound for the first conjugate point of $x = 1$ for the differential equation

$$x^2 y'' + xy' + (x^2 - \tfrac{1}{4})y = 0.$$

Answers

2. $c = 1.38$, approx.

3. $c = 2.36$, approx.

5. $c = 1 + \pi$.

4 Focal points

Consider again the functional

(4.1)
$$J = \int_a^b [r(x)y'^2 - p(x)y^2] \, dx,$$

where $r(x) > 0$, and $r(x)$ and $p(x)$ are continuous on an open interval that contains the closed interval $[a, b]$. With J we again associate the differential equation

(4.2)
$$[r(x)y']' + p(x)y = 0.$$

Perhaps the simplest *variable end point problem* in the calculus of variations is the following: Curves of class D' that join the point $(b, 0)$ to the line $x = a$ are termed *admissible*. Is there an admissible curve that provides a minimum value to J (see Fig. 10.3)?

In the earlier problem that we considered we found that the role of the conjugate point was definitive. The present problem is called a variable end point problem because it is required only that the curve join some point of the line $x = a$ with the point $(b, 0)$. For this problem the concept of focal point is crucial.

Let $z(x)$ be any nonnull solution of the differential equation (4.2) with the property that $z'(a) = 0$. If this solution has a first zero $x = f$ following $x = a$, the point $(f, 0)$ is called the *focal point* of the line $x = a$. It is easily seen that if any solution $z(x)$ has the property that $z'(a) = 0$, $z(f) = 0$, $z(x) \neq 0$ on $[a, f]$, all nonnull solutions having zero slope at $x = a$ have a first zero following $x = a$ at $x = f$.

The following theorem is fundamental.

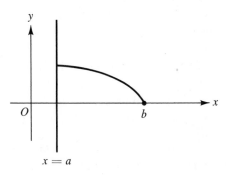

FIG. 10.3

Theorem 4.1. If the line $x = a$ has no focal point on the interval $[a, b]$, $J > 0$ for all admissible functions except along $y(x) \equiv 0$ for which $J = 0$. If the line $x = a$ has a focal point $x = f < b$, J can have no minimum in the class of admissible functions. Finally, if the first focal point $x = f$ of the line $x = a$ coincides with the point $x = b$, $J > 0$ except along solutions $cz(x)$, where $z(x)$ is a solution defining the focal point. Along these solutions, $J = 0$.

To prove the theorem, suppose first that the interval $[a, b]$ does not contain a focal point of the line $x = a$. Let $y(x)$ be an arbitrary admissible function, and let $z(x)$ be the solution defined by the conditions

$$z(a) = 1, \qquad z'(a) = 0.$$

This solution is positive on the interval $[a, b]$. Consider the equation (2.6)

$$J(y) = \left[r \frac{z'}{z} y^2 \right]_a^b + \int_a^b r \left(y' - \frac{z'}{z} y \right)^2 dx$$

(4.3)
$$= \int_a^b r \left(y' - \frac{z'}{z} y \right)^2 dx.$$

Thus $J(y) > 0$ for every admissible function $y(x)$ unless

(4.4)
$$y'(x) - \frac{z'(x)}{z(x)} y(x) \equiv 0,$$

that is, unless $y(x) \equiv cz(x)$, where c is a constant. Inasmuch as $y(b) = 0$, $z(b) > 0$, it follows that $c = 0$. Thus the function $y(x) \equiv 0$ is the only admissible function for which $J = 0$; for all others, $J > 0$.

Let us consider next the case when $x = b$ is the focal point of the line $x = a$. Let $z(x)$ be a solution of the differential equation that defines the focal point of $x = a$, let $y(x)$ be an arbitrary admissible function, and consider the equation (4.3). Note that y/z has a finite limit, as $x \to b$. Then,

$$J(y) = \int_a^b r\left(y' - \frac{z'}{z}y\right)^2 dx.$$

It follows that $J > 0$ along an admissible curve $y = y(x)$ unless (4.4) holds, that is, unless $y(x) \equiv cz(x)$. Conversely, if $y(x) = cz(x)$, it follows from equation (2.10) that $J = 0$.

It remains to be proved that a necessary condition that a minimum of J exists is that the interval (a, b) not contain a focal point of the line $x = a$. Suppose then that the line $x = a$ has a focal point $x = f(a < f < b)$. As in Section 1, it is easy to see that if J has a minimum, when evaluated among admissible functions, that minimum must be zero. It will then be sufficient to exhibit an admissible function for which $J < 0$.

To do this let $y(x)$ be the solution determined by the conditions

$$y(a) = 1, \qquad y'(a) = 0.$$

Then $y(f) = 0$, and $y(x) > 0$ on $[a, f)$, where $x = f$ is the focal point of the line $x = a$. Next, choose numbers h and d such that $a < h < f < d < b$, with d small enough that there is no point conjugate to $x = a$ on $[a, d]$. Then the points $[h, y(h)]$ and $(d, 0)$ can be connected by a unique solution curve $y = z(x)$ of the differential equation. Inasmuch as $y(d) \neq 0$, the solutions $y(x)$ and $z(x)$ are linearly independent. Abel's formula becomes

$$\Delta(x) = r(x)[y(x)z'(x) - z(x)y'(x)] \equiv c,$$

where c is a constant. But then

$$\Delta(f) = r(f)[-z(f)y'(f)] = c > 0.$$

Accordingly, $\Delta(h) > 0$; that is,

$$r(h)y(h)[z'(h) - y'(h)] > 0.$$

It follows that

(4.5) $z'(h) - y'(h) > 0.$

Define an admissible curve by the equations (see Fig. 10.4)

(4.6) $$y = \begin{cases} y(x) & (a \leq x < h), \\ z(x) & (h \leq x < d), \\ 0 & (d \leq x \leq b). \end{cases}$$

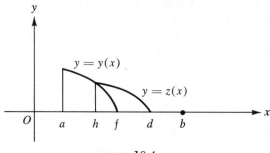

FIG. 10.4

We employ equation (2.10) and have

$$J(y) = [ryy']_a^h + [rzz']_h^d$$
$$= r(h)y(h)[y'(h) - z'(h)] < 0,$$

by (4.5).

The proof of the theorem is complete.

An upper bound for $x = f$, when the focal point exists, may be found in much the same fashion as were bounds for conjugate points earlier in this chapter. Consider, for example, the integral

(4.7)
$$J = \int_a^b (y'^2 - y^2)\, dx.$$

The focal point of the y-axis is, of course, the point $(\pi/2, 0)$. The line $y = b - x$ passes through the point $(b, 0)$. If we evaluate J along this curve and set the result equal to zero, we have

$$\int_0^b [1 - (b - x)^2]\, dx = 0,$$

or

$$3b - b^3 = 0.$$

It follows that $b = \sqrt{3} = 1.732$, approximately. The actual value of f is $\pi/2 = 1.57$, approximately, and $f < b$. An argument similar to that employed earlier makes it clear that such a number b is, indeed, an upper bound for f. In this case, the bound is not particularly close. Of course, functions other than $b - x$ may be employed.

The changes in the analysis required when the point is $(a, 0)$ and the line is $x = b$ will be clear.

Critique.   Theorems 2.1 and 4.1 remain valid as stated when the admissible functions are of class C' rather than of class D'. The proofs remain the same except for those portions of each that show that J cannot attain a minimum when a conjugate point, or focal point, lies on the open interval (a, b).

One approach is to round off the corners in (2.9) and in (4.6) "in a C' manner." This can be done rigorously, although handwaving is simpler—and, in this instance, reasonably satisfactory. Direct construction of a C' function for which $J < 0$ is possible, at least when b is not too much greater than, say, the conjugate point c. The method will be illustrated in the following example.

Consider

$$J = \int_0^b (y'^2 - y^2)\, dx,$$

$$y(a) = y(b) = 0,$$

where $b > \pi$. The function $\sin(\pi x/b)$ is of class C' on $[0, b]$ and vanishes at $x = 0$ and $x = b$. For this function.

$$J = \int_0^b \left[\frac{\pi^2}{b^2} \cos^2\left(\frac{\pi}{b}x\right) - \sin^2\left(\frac{\pi}{b}x\right) \right] dx$$

$$= \frac{b}{\pi} \int_0^\pi \left(\frac{\pi^2}{b^2} \cos^2 t - \sin^2 t \right) dt$$

$$= \frac{b}{2} \left(\frac{\pi^2}{b^2} - 1 \right) < 0,$$

since $\pi < b$.

In this example, b could have been any number larger than π. The student will discover that the method is valid for the more general functional

$$J = \int_a^b (ry'^2 - py^2)\, dx$$

provided b is not too much larger than the first conjugate point c of a.

Exercises

1.   Show that if there exists a solution $z(x)$ of (4.2) such that $z'(a) = 0, z(f) = 0$

$(a < f)$, $z(x) \neq 0$ on $[a, f)$, then every nonnull solution that has zero slope at $x = a$ has $x = f$ as its first zero following $x = a$.

2. Verify two statements in the last section of the proof of Theorem 4.1:
    (a) that $[h, y(h)]$ and $[b, 0]$ can be joined by a unique solution of the differential equation (4.2);
    (b) that $y(d) \neq 0$.

3. Find an upper bound of the focal point of the line $x = 0$ for the functional (4.7). Use $y = x^2 - b^2$.

4. Find an upper bound of the focal point of the line $x = 0$ for the functional

$$\int_0^b (y'^2 - xy^2) \, dx.$$

5. Find an upper bound of the focal point of the line $x = \pi/2$, when the differential equation (4.2) becomes

$$x^2 y'' + xy' + (x^2 - \tfrac{1}{4})y = 0.$$

6. Find the first focal point *less* than e^3 when the line is $x = e^3$, and the differential equation is

$$x^2 y'' + xy' + y = 0.$$

Answers

3. 1.58.

5. $f = 3.45$, approx.

6. $f = e^{\pi - 3}$.

Supplementary reading

There is a vast literature in ordinary differential equations, and it is difficult to pick a limited list of books for supplementary reading. Reading in any of the books that follow will prove rewarding to the student. It should be understood that this could be said as well of a list several times as long.

Barrett, J. H., *Oscillation Theory of Ordinary Differential Equations*, Advances in Mathematics, 3 (1969), pp. 415–508.

Coddington, Earl A., and Norman Levinson, *Theory of Ordinary Differential Equations*, McGraw-Hill, New York (1955).

Coppel, W. A., *Disconjugacy*, Springer-Verlag, Berlin (1971).

Corduneanu, C., *Principles of Differential and Integral Equations*, Allyn and Bacon, Boston (1971).

Hale, Jack K., *Ordinary Differential Equations*, Wiley-Interscience, New York (1969).

Hartman, Philip, *Ordinary Differential Equations*, Wiley, New York (1964).

Hurewicz, W., *Lectures on Ordinary Differential Equations*, M.I.T., Cambridge, Mass. (1958).

Krasovskii, N. N., *Stability of Motion* (translated by J. L. Brenner) Stanford, Stanford, Calif. (1963).

LaSalle, Joseph, and Solomon Lefschetz, *Stability by Liapunov's Direct Method, with Applications*, Academic, New York (1961).

Morse, Marston, *Variational Analysis*, Wiley, New York (1973).

Reid, William T., *Ordinary Differential Equations*, Wiley-Interscience, New York (1971).

Roxin, Emilio O., *Ordinary Differential Equations*, Wadsworth, Belmont, Calif. (1972).

Sanchez, David A., *Ordinary Differential Equations and Stability Theory: An Introduction*, Freeman, San Francisco (1968).

Sansone, G., and R. Conti, *Non-linear Differential Equations*, Pergamon, Macmillan, New York (1964).

Swanson, C. A., *Comparison and Oscillation Theory of Linear Differential Equations*, Academic, New York (1968).

Wong, Pui-kei, *Sturmian Theory of Ordinary and Partial Differential Equations*, National Taiwan University (1971).

Index

247